GENIUSES, HEROES, AND SAINTS

GENIUSES, HEROES, AND SAINTS

THE NOBEL PRIZE AND THE PUBLIC
IMAGE OF SCIENCE

MASSIMIANO BUCCHI

TRANSLATED BY TANIA ARAGONA

THE MIT PRESS   CAMBRIDGE, MASSACHUSETTS   LONDON, ENGLAND

The MIT Press
Massachusetts Institute of Technology
77 Massachusetts Avenue, Cambridge, MA 02139
mitpress.mit.edu

© 2025 Massachusetts Institute of Technology

Originally published as *Come vincere un Nobel: Il premio più famoso della scienza*, © 2017, Giulio Einaudi editore s.p.a., Torino. https://www.einaudi.it

The translation of this work has been funded by SEPS (Segretariato Europeo per le pubblicazioni scientifiche)—www.seps.it, seps@seps.it.

All rights reserved. No part of this book may be used to train artificial intelligence systems or reproduced in any form by any electronic or mechanical means (including photocopying, recording, or information storage and retrieval) without permission in writing from the publisher.

This book was set in Stone Serif and Avenir LT Std by Westchester Publishing. Printed and bound in the United States of America.

Library of Congress Cataloging-in-Publication Data

Names: Bucchi, Massimiano, 1970– author | Aragona, Tania translator
Title: Geniuses, heroes, and saints : the Nobel Prize and the public image of science / Massimiano Bucchi ; translated by Tania Aragona.
Other titles: Come vincere un Nobel. English | Nobel Prize and the public image of science
Description: Cambridge, Massachusetts : The MIT Press, [2025] | "Originally published as Come vincere un Nobel, © 2017, Giulio Einaudi Editore S.p.A., Torino."—t.p. Verso. | Includes bibliographical references and index.
Identifiers: LCCN 2024023053 (print) | LCCN 2024023054 (ebook) | ISBN 9780262551847 paperback | ISBN 9780262382526 epub | ISBN 9780262382533 pdf
Subjects: LCSH: Nobel Prizes | Science—Awards | Nobel Prize winners
Classification: LCC AS911.N9 B8313 2025 (print) | LCC AS911.N9 (ebook) | DDC 001.4/4—dc23/eng/20250108
LC record available at https://lccn.loc.gov/2024023053
LC ebook record available at https://lccn.loc.gov/2024023054

ISBN: 978-0-262-55184-7

10 9 8 7 6 5 4 3 2 1

EU product safety and compliance information contact is: mitp-eu-gpsr@mit.edu

CONTENTS

ACKNOWLEDGMENTS vii

1 HOW A MISTAKEN OBITUARY LED TO THE MOST IMPORTANT AWARD IN SCIENCE 1
2 HOW DO YOU WIN A NOBEL PRIZE? 17
3 A PRIZE WITH NO BORDERS? NO POLITICS PLEASE, WE ARE SWEDISH (SCIENTISTS) 35
4 AND THE WINNER IS . . . NOT HERE! HOW EINSTEIN WON THE NOBEL PRIZE AND WHY HE ALMOST NEVER RECEIVED IT 57
5 HOW NOT TO WIN A NOBEL PRIZE: THE STORY OF LISE AND OTHER PRIZE GHOSTS 79
6 DOES THE PRIZE MAKE THEM MORE APPEALING? THE IMPORTANCE OF BEING A NOBEL 103
7 THE BODY OF THE (LAUREATE) SCIENTIST 133
EPILOGUE: GENIUSES, HEROES, AND SAINTS—HOW THE NOBEL PRIZE (RE)INVENTED THE PUBLIC IMAGE OF SCIENCE 159

APPENDIX: ALL THE NOBEL LAUREATES IN THE SCIENCES, 1901 TO 2024 163
NOTES 173
INDEX 189

ACKNOWLEDGMENTS

The idea of studying the Nobel Prize and its effects on the public image of science dates back to a study visit in Sweden in 1998. There are many colleagues, friends, and institutions I need to say thanks to for their help over the past twenty years:

To director Karl Grandin and all the staff of the Center for the History of Science at the Royal Swedish Academy of Sciences for giving me the opportunity to conduct my research in the archive of the decisions relating to the prizes.

To the Italian Cultural Institute of Stockholm for its hospitality and support through the Lerici program.

To everyone at the Nobel Museum and Library, in particular Erika Lanner and Ulf Larsson.

To Anders Bárány (former secretary of the Nobel Committee for Physics) and Erling Norrby (former secretary of the Royal Swedish Academy of Sciences); Aant Elzinga, Gustav Källstrand, Nils Hansson, Svante Lindqvist, Katarina Nordqvist, Adrian Thomasson, Sven Widmalm, and all the other colleagues with whom I had the pleasure of discussing these topics during seminars, conventions, and informal conversations at the University of Uppsala, the Royal Swedish Academy of Sciences, the Nobel Prize Museum, and the University of Düsseldorf.

To Lorenzo Beltrame and Silvia Giovannetti for their invaluable help in collecting and analyzing articles in the press on the Nobel Prize.

To Marco Beretta and Luca Critelli for their help in sourcing some fundamental texts for this work.

To the staff of the Widener Library, the Houghton Library, and the Harvard University Archives for their assistance and advice.

To David Kaiser of the Massachusetts Institute of Technology; to Steven Shapin, Sheila Jasanoff, and the students of the Program on Science, Technology, and Society at the Kennedy School of Government (Harvard University) for interesting discussions on these themes; to John Durant (MIT) for a fundamental piece of information on Salvatore Luria; to Felice Frankel (MIT) and Diane Booton for their help during my time in Boston.

To Ayumi Koso and Motoko Kakubayashi at the University of Tokyo for hosting my seminar and providing information on how the Nobel Prizes and the staff of the relative institutions manage the first hours following the announcement of the prize from a communications point of view.

To the Nobel Prize laureates who patiently allowed me to interview them over the past years, in particular Wolfgang Ketterle (MIT) and Hiroshi Amano (University of Nagoya).

I am grateful to Marco Cavalli and Renato G. Mazzolini for their careful reading and many suggestions; to Alberto Brodesco, Giuseppe Pellegrini, Barbara Saracino, and Vincenzo Barone for their comments on the preliminary version of this text; to Eliana Fattorini for her valuable help in proofreading.

1

HOW A MISTAKEN OBITUARY LED TO THE MOST IMPORTANT AWARD IN SCIENCE

In less than five minutes I shall have thrown my pen into the fire. . . . I have but half a score things to do in the time—I have a thing to name—a thing to lament—a thing to hope, a thing to promise, and a thing to threaten.
—L. Sterne, *The Life and Opinions of Tristram Shandy, Gentleman*, 1759

A MISTAKEN OBITUARY

Imagine the scene. In April 1888, Alfred Nobel has just awoken in his beautiful home in Paris at number 55 avenue de Malakoff. He skims through the newspapers as he eats his breakfast. Suddenly, he jumps out of his chair. There, in the paper, he sees his obituary. Yes, it is definitely his! And what a title it has: "The Merchant of Death Is Dead." "Alfred Nobel, who made his fortune by discovering the means of killing people in the quickest way possible, passed away yesterday," said the newspaper.

Incredulous, Alfred reads the announcement again, then shakes his head and smiles sadly. He realizes that the paper has mistaken Alfred for his older brother Ludvig, who died in Cannes a few days earlier. Alfred gets up from the table, his appetite lost, but sits back down and picks up the newspaper again. The name is wrong, the obituary is wrong, but that title, "The Merchant of Death Is Dead," was intended for him: a chemist, inventor, and entrepreneur of great success, the holder of 355 patents, among them those for dynamite and explosive gelatine, although his inventions had almost never been used in the context of war.

So this is how I will be remembered, he realizes sadly. Alfred then returns to his laboratory, to his thousands of projects and his mostly solitary life. But the thought of that obituary and the harsh judgment from his peers continues to nag him.

The death of his brother Ludvig was not the first tragedy for Alfred and his family. In September 1864, the small nitroglycerin factory that he had set up in Heleneborg, south of Stockholm, blew up, killing his twenty-one-year-old brother, Emil Oskar. The event worsened Alfred's father's health, and he died in 1872, followed the next year by Alfred's mother.

Born in Stockholm on October 21, 1833, and raised in Russia after the age of nine, Alfred was a true cosmopolitan. He often traveled and learned five languages fluently at a very young age. In the mid-1860s, he opened a factory in Germany, south of Hamburg. His explosives were efficient but highly dangerous. After a series of accidents, nitroglycerin was eventually banned in some countries. In order to make nitroglycerin safer to handle, it was necessary to find a porous material and other substances with which to mix it.

Alfred undertook many experiments to no avail. But while walking near the German factory, he discovered a porous rock, called *kieselguhr* or diatomite, which, when mixed with nitroglycerin, made it possible to mold and, above all, dilute it, thus making it possible to control its explosive potential. Alfred called the compound *dynamite*, from the Greek word for "power." He patented it in 1867 across Europe and was immediately inundated by orders. The turning point for his success came in 1870, when dynamite was the decisive element for the construction of the Gotthard Tunnel. Over the course of his career, Alfred opened approximately ninety factories in twenty different countries.

In 1868, Alfred and his father received a prize from the Royal Swedish Academy of Sciences: the Letterstedt Prize, awarded for "important discoveries of practical value for humanity." Alfred might have had that prize and its motivation in mind a few years later when he wrote his own will.

In 1875, Alfred bought a house in Paris, on avenue de Malakoff, where he set up a laboratory, which he subsequently transferred to Sevran when it became too small. In 1891, following a series of disagreements with the

French authorities, he decided to buy a villa in Sanremo in Italy, where he also set up a laboratory (which he confidentially called "my nest"). In the meantime, he acquired another house in Bofors, Sweden.

When he was not traveling, Alfred led a solitary life, focusing mainly on his work. Among his hobbies was a passion for literature, especially poetry. "A recluse with no books or ink is a living dead man," he would say. Alfred interpreted the poem "Prometheus Unbound" by his favorite poet, Percy Bysshe Shelley, as an allegory of the extraordinary forces liberated by science and technology during his lifetime and perhaps even of his own experience with explosives. He also liked to write: as a young man, he wrote poems, mostly in English; he started a few tentative novels, mostly satirical works inspired by his experiences as an inventor, including a comedy called *The Patent Bacillus*.

Alfred assiduously avoided the celebrity that inevitably came with his inventions and economic success. "I am not aware that I have deserved any notoriety and have no taste for its noise," he said.[1] He refused to be interviewed or have his portrait taken. His most famous portrait, a painting that now hangs at the Nobel Foundation in which he is sitting in front of his test tubes and alembics while staring out melancholically into space, was painted almost twenty years after his death. When his brother Ludvig requested a biography, he replied self-deprecatingly, referring to his lifelong poor health, "Alfred Nobel, a miserable half-life, ought to have been choked to death by a philanthropic physician as soon as, with a howl, he entered life."

Ten years before the mistaken obituary, Alfred himself directed a newspaper to print another curious message. In 1876, an Austrian paper published the following: "a wealthy, highly educated, elderly gentleman, living in Paris, is looking for a lady of mature years with a knowledge of languages to act as his secretary and housekeeper."[2]

The Austrian countess Bertha Kinsky von Wchinitz und Tettau replied to the advertisement. She was thirty-three, of a noble family that had fallen in disgrace. Nobel employed her immediately, perhaps seduced by her aristocratic beauty. But one day, he found a letter from her on his desk. The countess had sold her jewels to buy a ticket back to Vienna, where she met up with Arthur von Suttner to elope secretly from their respective families.

The event was a blow for Alfred. The following year, he had an affair with the young Viennese Sofie Hess. Alfred took Sofie to Paris, but the two soon realized they did not have much in common. Nobel bought her a house in Austria, maintained her high living standards, and even encouraged her to marry a military officer. For his part, Alfred resigned himself to a solitary life.

Nevertheless, he wrote regularly to Bertha, although they did not see each other for the next eleven years. In 1889, Bertha became one of the main figures of the pacifist movement thanks to the international success of her book *Lay Down Your Arms*. Nobel liked to listen to Bertha's ideas, although he never refrained from expressing skepticism. When she invited him to take part in a pacifist congress, he replied,

My factories may make an end of war sooner than your congresses. The day when two army corps can annihilate each other in one second, all civilized nations, it is to be hoped, will recoil from war and discharge their troops.[3]

In a letter dated January 7, 1893, Nobel announced to Bertha his intention to create a prize for peace:

I am prepared to set aside a part of my estate for a prize to be awarded every fifth year—let us say six times, because, if in thirty years it has not been possible to reform the present system, we shall unavoidably fall back into barbarism. This prize would be awarded to the man or woman who had induced Europe to take the first step toward the general idea of peace.[4]

HISTORY'S MOST FAMOUS WILL

Early in December 1895, four men received a curt invitation to the Swedish-Norwegian Club in Paris. When they arrived, the four—Thorsten Nordenfelt, a war industrialist; Sigurd Ehrenborg, a retired officer of the Swedish army; and two engineers, R. W. Strehlenert and Leonard Hwass—understood the reason for the invitation.

Their friend Alfred Nobel wanted them to witness him signing his will, which he drew up a few days earlier in his house in Paris without consulting a lawyer. The will was brief, contained on a single page, and canceled all his previous dispositions.

Only a few lines are of interest to us, but they would change science and its public image forever:[5]

All of my remaining realizable assets are to be disbursed as follows: the capital, converted to safe securities by my executors, is to constitute a fund, the interest on which is to be distributed annually as prizes to those who, during the preceding year, have conferred the greatest benefit to humankind. The interest is to be divided into five equal parts and distributed as follows: one part to the person who made the most important discovery or invention in the field of physics; one part to the person who made the most important chemical discovery or improvement; one part to the person who made the most important discovery within the domain of physiology or medicine; one part to the person who, in the field of literature, produced the most outstanding work in an idealistic direction; and one part to the person who has done the most or best to advance fellowship among nations, the abolition or reduction of standing armies, and the establishment and promotion of peace congresses. The prizes for physics and chemistry are to be awarded by the Swedish Academy of Sciences; that for physiological or medical achievements by the Karolinska Institute in Stockholm; that for literature by the Academy in Stockholm; and that for champions of peace by a committee of five persons to be selected by the Norwegian Storting. It is my express wish that when awarding the prizes, no consideration be given to nationality, but that the prize be awarded to the worthiest person, whether or not they are Scandinavian.
—Alfred Bernhard Nobel, Paris, November 27, 1895

This short text contains everything that was important to Alfred Nobel. Science was his work and his life, but he was also passionate about literature, a great source of satisfaction. World peace became a passion of his during his final years. And he had studied physics and chemistry since childhood, with the help of two eminent university professors, Yuli Trapp and Nikolaj Zinin. Finally, his life as a globetrotter and polyglot led him to underscore his "wish that when awarding the prizes, no consideration be given to nationality," a point that would turn out to be crucial.

Alfred Nobel died in Sanremo on December 10, 1896, succumbing to a brain hemorrhage.

Until a few days after his funeral, not even his closest family members knew anything about his will, which began the story of the most important prize in the scientific world: the most famous invention by a man who had 355 patents. With a few lines, Alfred Nobel made history. The name of a man who hated any form of publicity or fame during his lifetime became known to everyone after his death.

THE YOUNG ENGINEER WHO DEFENDED THE WILL AT GUNPOINT

On the evening of December 15, 1896, the twenty-six-year-old engineer Ragnar Sohlman is in bed in his hotel room in Sanremo, where he has just arrived for Alfred's funeral. Sohlman had been Nobel's personal assistant during the last years of his life. He awakens to a knock on the door and finds himself face to face with Emanuel and Hjalmar Nobel, Alfred's nephews. They have just received a telegram informing them of the existence of Alfred's will and that Sohlman and the industrialist Rudolf Lilljequist are to be its executors. Sohlman pales and falls back onto the bed. Emanuel also blanches when he finally reads the will and compares it to the previous versions. Alfred has substantially reduced his inheritance and that of other family members in favor of marking assets for the prize. Furthermore, Emanuel has been replaced as executor by Sohlman and Lilljequist.

Sohlman's worries are well justified. He now faces a race against time to execute Alfred's last wishes. The Nobel Foundation, designated as the primary beneficiary, for example, does not yet exist: How and by whom will it be created? Under which jurisdiction should the will be executed, since during his last years Alfred lived mostly between Paris and Sanremo, although he specified Swedish and Norwegian institutions should be responsible for handing out the prizes? And how can Sohlman defend the will from the judicial attacks that will inevitably be set in motion by Alfred's family?

On January 2, 1897, the Swedish public and afterwards the world were informed of the odd will. The initial surprise was soon followed by a wave of controversy. Swedish opinion claimed that the will was unpatriotic, because it took into account foreign beneficiaries; that it created insurmountable practical issues; and that it exposed the members of the designated prize-awarding institutions to the risks of corruption. The choice of the Norwegian parliament for the assignation of the Peace Prize also caused controversy, although Sweden and Norway had been united since 1814.

The head of the Swedish Social Democratic Party, Hjalmar Branting, wrote a furious article titled "The Testament of Alfred Nobel: Noble Intentions, Extraordinary Mistake." It criticized Nobel for not having "shown any interest for social well-being" during his lifetime, as well as the nature of the prize (what was the point of awarding prizes to well-established scholars?) and the institutions designated for choosing the award recipients. According to Branting, it was the "working-class masses," not individuals, who were capable of achieving world peace, and therefore the masses deserved part of the Nobel Foundation's money: "It would be better—concluded Branting dramatically—to be rid of both millionaires and donations."[6]

Sohlman and Lilljequist embark on a race against time. Their first objective is to retrieve all of Nobel's assets from France in order to protect them from France's inheritance laws and any legal action, such as seizure, by Alfred's family. Every day for a week, the two men withdraw two and a half million Francs in stocks from the coffers of the Comptoir national d'escompte de Paris and drop them off at the Swedish consulate. Sohlman guards the money in transit, a loaded gun in his hand. At the consulate, the assets are wrapped and sent to London (the part to be sold) and to Stockholm (the part to be saved). During these extraordinary maneuvers, Nobel's family members occasionally pay visits to the consulate to plead their case; at times they do so while the two executors are wrapping up Alfred's assets in the next room.

By the end of this meticulous operation, after traversing all of Europe, Sohlman and Lilljequist have amassed approximately 33 million kronor. After taxes, they have around 31 million kronor (approximately 190 million euros as of 2024) with which to execute Alfred's last wishes. But they still have to reach an agreement with Nobel's young lover, Sofie Hess. Unsatisfied with the amount that Alfred has set aside for her in his will, Sofie threatens to make public their potentially embarrassing correspondence.

Besides settling with Hess, they must have Alfred's will validated in Sweden, also in order to overcome the difficulties that the few lines signed by Alfred presented with regard to French legislation. The task was anything but straightforward: Nobel had not been a permanent resident in Sweden since he was nine, although he had bought a house there in his final years.

In May 1897, following the testimony of witnesses of Alfred's last wishes before the court in Stockholm, a government decree ordered the Swedish general prosecutor to validate the will. The government's justification for intervening to "facilitate the noble intentions of the testator" rested on Nobel's Swedish citizenship, the involvement of Swedish institutions (such as the Swedish Academy, the Royal Swedish Academy of Sciences, and the Karolinska Institute), and the fact that the donation was in the interest of the general public.[7]

THE SWEDISH SCIENTISTS WHO DID NOT WANT THE NOBEL PRIZE

The third obstacle, which the two executors of the will perhaps did not anticipate, is presented by the same scientific institutions that Alfred had appointed to assign the science prizes—the Royal Swedish Academy of Sciences and the Karolinska Institute.[8] Some members of the institutions worried that allocating prizes might lead to internal divisions and would have preferred to dispose of a slice of the assets left by Nobel for the institutions' own objectives. Sohlman, however, was firm on this point: the dispositions stated in the will were clear and had to be respected. Others contested the idea of the prize itself. For the chemist Otto Pettersson, the idea of putting scientists in competition against one another was "the stupidest use of a bequest that I can imagine! To seek reward for their work is not attractive for scientists."[9]

A third faction stressed the risk of finding themselves in a never-ending judicial controversy. This was the argument that professor Hans Forssell, president of the Swedish Academy and member of the Academy of Sciences, intent from the start on boycotting the initiative, put forward to convince his colleagues to turn down the proposal.

The executors thus suddenly find themselves at a dead end. But they push forward with a series of informal meetings with the Swedish academics. How can they carry out Alfred's wishes, establish a functioning system to distribute prizes, and respect the independence of the institutions concerned?

The eventual involvement of Emanuel Nobel proves key to reconciling all parties. Though disappointed by the contents of the will, Emanuel expresses his intention to carry out his uncle's wishes. He gives his approval after a negotiation that leads to a financial agreement between him and other members of the family, allowing them to acquire the entire package of stocks of the Nobel Brothers Petroleum Company based in Baku. The company produced up to 50 percent of the petroleum sold worldwide, providing a degree of influence that gave the Nobel family the nickname "Rockefellers of Russia." Emanuel goes as far as defending his uncle's will in a conversation with the king of Sweden, Oscar. When the king reminds him that "it is your duty towards your brothers and sisters, who are your wards, to ensure that their interests are not neglected to the advantage of your uncle's extravagant ideas," Emanuel answers with such forcefulness ("Your Majesty, I would not care to expose my sisters and brothers to the risk of being reproached, in the future, by distinguished scientists for having appropriated funds which properly belong to them")[10] that his Russian lawyer advises him to leave Sweden in order to avoid being arrested for lèse-majesté. A financial contribution is guaranteed to the institutions that are to award the prizes to help them to select candidates.

Finally, on June 29, 1900, the Swedish government ratifies the Statutes of the Nobel Foundation. The statutes allow for the possibility of awarding a prize to two works within the same field or to two or three persons having collaborated on the same work. However, "in no case may a prize amount be divided between more than three persons."[11]

Furthermore, in order to allow the scientific institutions more freedom, the phrase "during the previous year" is interpreted broadly. In particular, "the most recent achievements in the fields of culture referred to in the will and for older works only if their significance has not become apparent until recently." If no work is considered prizeworthy in one year, the purse can be put aside for the following year.

The statutes also require that each institution set up a three- to five-member committee for the preliminary examination of the proposed candidates. The Nobel Prizes, once awarded, are to be irrevocable, and the committee's deliberations are to be classified and not to be appealed. The institutions can release relevant documentation of a prize only fifty years after it has been awarded.

Finally, it is established that the prize be awarded on the anniversary of Alfred Nobel's death on December 10. Before the prize giving or within six months of the prize being awarded, "if possible," the laureate is required to hold a public lecture in Stockholm or, in the case of the Peace Prize, in Oslo.

DECEMBER 10, FIVE YEARS LATER, IN STOCKHOLM

Sohlman's frenetic obstacle course concludes on December 10, 1901, on the fifth anniversary of Alfred Nobel's death.

In the reception room of the Royal Swedish Academy of Music, decorated for the occasion by the royal architect Agi Lindegren, the secretary of the Nobel Foundation and Prime Minister Bostrom raise their glasses to the memory and wishes of Alfred Nobel. The Secretary of the Academy of Sciences then introduces the first Nobel Prize winners in history.

The first Nobel Prize for physics is awarded to the German Wilhelm Conrad Röntgen, "in recognition of the extraordinary services he has rendered by the discovery of the remarkable rays subsequently named after him" (X-rays). The Nobel Prize for chemistry is awarded to the Dutch Jacobus Henricus van 't Hoff "in recognition of the extraordinary services he has rendered by the discovery of the laws of chemical dynamics and osmotic pressure in solutions." Finally, the president of the Karolinska Institute introduces the German Emil von Behring, awarded for physiology or medicine,

for his work on serum therapy, especially its application against diphtheria, by which he has opened a new road in the domain of medical science and thereby placed in the hands of the physician a victorious weapon against illness and deaths.

In his absence due to an illness, the Nobel Prize for literature awarded to Sully Prudhomme is handed to the French ambassador. Since King Oscar was away on a diplomatic visit, Prince Gustav bestows a diploma and medal, along with a substantial prize of 150,782 kronor (approximately 970,000 euros as of 2024) to each recipient. (To put this into context, the money was the equivalent of the total sum of all prizes handed out between 1901 and 1910 by the French Academy of Sciences.) At the end of the ceremony, to the notes of a march played by the Royal Orchestra, the royal family, the laureates, and guests make their way to the Grand Hotel for a celebration banquet.

The menu includes starters, "turbot supreme à la normande," "filet à l'impériale," game, salad, and various pastries. They are served an 1897 Nierstener and an 1881 Château Abbé Gorsse and raise their glasses with a Crème de Bouzy champagne. When his turn comes for a speech, Röntgen chooses to talk about northern mythologies, drawing a comparison with his own adventure, which came to life with the discovery of X-rays.

The international press was immediately impressed by the ceremonial aspect of the proceedings:

> It is impossible to send you a detailed account by telegraph of the speeches, the choirs, the different types of music and the grand banquet that followed the proclamation by the Swedish Academy of the Nobel Prize winners. The ceremony ended late. The winners were called in the following order: Roentgen for physics, Vant Hoff [sic] for chemistry, Behring [sic] for physiology.[12]

As confirmed by the short description published by the *Corriere della Sera*, Röntgen's discovery is by far the most well-known to the public (see chapter 6):

> It is needless to mention the numerous practical applications of Röntgen's rays. They are not just an advantage for physics but also for medicine and surgery.[13]

However, there was criticism too. The press expressed a few doubts on the choice of Von Behring. According to the *Corriere*,

> [. . .] the clear paternity of this discovery has already been contested by numerous French scientists, although he is without doubt one of Germany's greatest bacteriologists.

A group of writers, artists, and critics wrote a letter in protest to the Russian writer Lev Tolstoj [Leo Tolstoy]. In it, they expressed admiration for Tolstoj, who, according to them, should have been awarded the prize. They also gave a scathing judgment that the Academy "which has control over said prize reflects neither the view of the artists nor of public opinion."[14]

Among those who signed the letter were famous writers such as August Strindberg and two future Nobel Prizes for literature, Selma Lagerlöf and Verner von Heidenstam.

And this is how, with admirable determination, the young Sohlman carried out the task that Alfred imposed on him. Five years earlier, very few people would have bet on him to succeed in accomplishing this complicated mission—one hampered by legal and financial complications, the hostility of Nobel's family, and the divided opinions of academics and the public.

Nobel's surprising choice to entrust everything to Sohlman turned out to be the right one. The engineer defended his mentor's wishes against all critics without ever questioning them. Nevertheless, he may not have succeeded without the farsightedness of the Swedish government, the willingness of Emanuel Nobel to accept his uncle's decisions, and the understanding of some academics that the prize would give international visibility to their institutions.

Bertha von Suttner, to whom Nobel had written when first contemplating a prize for advancing peace, was the first woman to receive the Nobel Peace Prize in 1905. Ironically, one of the most vocal critics of the prize at the time, Hjalmar Branting, would also receive the Peace Prize in 1921.

A PRIZE THAT FASCINATED THE MEDIA AND THE PUBLIC

From the very beginning, the story of the Nobel Prize was linked to the public image of science.

The astounding sum of money accompanying each prize and the controversies that erupted following the publication of Alfred's will immediately attracted media attention.

No sooner had the will been made public than news of Nobel's legacy appeared in more than a hundred newspapers across the world. It was the first time that an industrialist's fortune was earmarked for financing prizes. Following the announcement of the first prize winners, over five hundred articles about Nobel's will were published. Between December 1901 and June 1902, over a thousand articles and comments on the prize and its founder appeared.[15]

The combination of such diverse fields (sciences, literature, and peace) turned out to be a strength. On one hand, the combination of peace and literature ("of idealistic inspiration," according to the founder's wishes) lent a spiritual aura to the scientific prizes. On the other hand, the association with the material achievements of scientific progress gave relevance and visibility to the prizes for peace and literature. These last two prizes were the ones that especially attracted the initial attention of public and press. But the science prizes were the ones that contributed particularly to the reputation and authoritativeness of the prize.

In 1904, the American newspaper the *Evening Transcript* wrote that

Thus far the credentials of all those who have been distinguished by the prizes are approved by the popular jury as much as the special one that made the award.

In 1906, *Cosmopolitan* asserted that

The history of modern science might be written without going outside of the names of the Nobel prizes for beneficient [sic] discoveries in physics, chemistry, and medicine.[16]

The secrecy of the selection process piqued the interest of journalists, who competed to preempt their colleagues in divulging the news, sometimes attempting to monitor the arrival of foreign scientists in Stockholm. This is probably how the *Corriere della Sera*, republishing an article from a French paper, correctly reported (albeit with a few spelling mistakes) the first science laureates on December 8, 1901 (though not the one for literature, which the paper mistakenly awarded to the Spaniard José Echegaray).

The most sensational instance of speculation during those first years attributed the 1908 Nobel Prize for physics to the German Max Planck, who, in fact, would not receive a Nobel until 1918. This contributed to the decision, in 1910, to announce the laureates immediately after the official decision instead of waiting until the December ceremony. However, as we will see, this did not put an end to all speculation and jockeying for information, as illustrated by the bizarre events of the non-awarding of the Nobel to Thomas Alva Edison and Nikola Tesla (see chapter 6).

The curiosity of the media and the public about scientific prizes was furthermore enhanced, as we shall see, by the charismatic figures of some laureates, such as Marie Curie (Nobel Prizes in 1903 and in 1911) and Guglielmo Marconi (1909).

The contradictions inherent in the prize particularly caught the public imagination: of its founder, who with the prize wished to balance the success and substance of his career with inventions that were as lucrative as they were controversial; and of science itself, whose power and practical implications, starting with Nobel's own work, were beginning to appear as obvious as its knowledge ambitions.

On December 13, 1901, noting the first prize-giving ceremony, the *Corriere della Sera* reported that

[Nobel] is famous for two reasons: for having invented dynamite and for having founded the prizes that carry his name when he died. The first presentation took place just a few days ago. It took five years for the legal and effective existence of his foundation to come to life, overcoming legal suits etc. From now on, every year, five prizes will be awarded—one for literature, one for physics, one for chemistry, one for medicine and physiology, and one for peace. This way, most of the immense fortune left by Nobel—approximately 50 million—will serve humanity's highest interests. [. . .] It was Nobel's wish to give to charity after his death in the same way he had done so during his life. He had warned his family members that he would leave them only memories. He scorned fortunes obtained through inheritance and respected only those acquired through hard work. But knowing that study and even the greatest of discoveries rarely enriched their authors, he left his wealth to men and women of letters, scientists, and philanthropists. The inventor of dynamite, having dedicated his mind to the art of warfare, had dedicated his legacy to the art of peace keeping.

OVER A CENTURY OF PRIZES (AND CONTROVERSIES)

The machine that Nobel's last wishes set in motion over a hundred years ago has (almost) never stopped. Between 1901 and 2024, 1007 persons and 31 organizations[17] were awarded the Nobel Prize. More specifically, for the sciences that we are focusing on here, 112 prizes have been awarded for physics to 210 scientists, 110 for chemistry to 181 scientists, and 109 prizes for medicine and physiology to 216 scientists.

In over a century, the prizes have not been without criticism and controversies. The most common ones concerned the choice of the recipients. In chapters 4, 5, and 6, we will look at some of the rather questionable awards (or that today are considered debatable) and at how some significant contributions and scientists are "ghosts" that have been ignored by the prize. We will also look at how one of the greatest and most celebrated scientists of the twentieth century was almost left out of the world's most famous prize for science.

Today it is the Nobel Prize itself that is frequently the object of criticism from scientists and commentators. The criticism underlines two shortcomings of the prize in particular. The first concerns the restriction to specific disciplines, which reflected the state of research (and of Alfred Nobel's personal experience) and its social relevance at the time of the prize's creation. The prize has not been extended to newly relevant areas of research, such as environmental science, nor does it recognize that scientific progress is increasingly interdisciplinary.[18]

A second limitation is the focus on individual scientists. Over the past fifty years, research has become an increasingly collective enterprise, with large research groups publishing articles signed by hundreds and even thousands of researchers. This is why in 2013, for example—when the Nobel Prize for physics was awarded to Peter Higgs and François Englert "for the theoretical discovery of a mechanism that contributes to our understanding of the origin of mass of subatomic particles, and which recently was confirmed through the discovery of the predicted fundamental particle [the so-called Higgs boson], by the ATLAS and CMS experiments at CERN"—there was much criticism for the failure to include the researchers from CERN among the recipients.[19]

According to David Kaiser, a physicist and historian of physics at the Massachusetts Institute of Technology, the prize is very much a product of its time, exemplified by the famous group photo at the Solvay Congress in 1927 that immortalized Einstein, Bohr, Marie Curie, and many other past or future Nobel Prizes:

That photo symbolizes the real scientific community of those times in every sense of the word: they all more or less knew each other and would all have been able to broadly explain the contributions of all their Nobel laureate colleagues. Since then, things have changed considerably, even from a quantitative point of view. Today, there are approximately one hundred thousand physicists in the United States alone and an increasing level of specialization and articulation in the various research fields. The result is that when awarding the prizes, there is more or less implicitly an alternation within the subsectors.[20]

It is difficult to say, however, if the Nobel Prize could effectively be rethought and, if so, how it could realistically comply with the criticism and changes. According to Kaiser,

in principle, to include new sectors, such as economics, is not impossible. And as far as the collective dimension is concerned, the prize for peace has often been awarded to groups and organizations. As for sciences, one could consider rewarding research groups while citing or interviewing their "spokesman" in a way that would reflect the manner in which major experiments are organized.

During the past few years, other prizes have been created based on these aspects in order to "compete" with the Nobel Prize. Some of the recently created ones, such as the Breakthrough Prizes founded by Russian millionaire Yuri Milner, have much greater funding ($3 million for each prize compared to the $980,000 for the Nobel) and cover sectors (such as neurosciences and mathematics) that are not covered by the Swedish prize or areas that are considered to be overlooked by the Nobel Prize (such as theoretical physics). These prizes have also begun to be awarded to entire research groups. In 2016, the Breakthrough was awarded to the recording of gravitational waves: $1 million was awarded to the three founders of the LIGO experiment, and $2 million was divided among the other 1,012 scientists who participated in the experiment.

The Nobel Prize is not without its shortcomings. However, it is possible that these "limitations"—specific disciplines and the focus on individuals—have enhanced the public image of the Nobel Prize, thus contributing through it to shape the public image of science and scientists during the course of the last century. We will look at this in the next chapters. Now it is time to find out how to win a Nobel Prize.

2

HOW DO YOU WIN A NOBEL PRIZE?

If I knew what leads one to the Nobel Prize, I wouldn't tell you, but go to get another one.
—R. B. Laughlin, Nobel Prize for Physics, 1998

So how does one win a Nobel Prize? Recall that Nobel named only three scientific subjects in his will: physics, chemistry, and physiology and medicine (these last subjects count as one, and the prize is called "Physiology or Medicine"). Mathematics, geology, botany, and astronomy are not mentioned.

One widespread rumor claims that Nobel excluded mathematics because he could not stand the Swedish mathematician Gösta Mittag-Leffler and was afraid that the prize might be awarded to him. Their animosity was said to date back to an old rivalry, which was either sentimental or financial; some even said that Mittag-Leffler had had a relationship with Nobel's wife (?). In reality, the two barely knew each other (and Nobel, as was mentioned, never married).[1] One possible explanation for this rumor is that it was a subsequent rationalization following a subject exclusion from the prize that was perceived as unfair and unmotivated.

In fact, as we have seen, Nobel focused on the subjects that he had studied when he was young and that he considered to have the most impact on the life and well-being of humankind.

Only in 1968 did the Bank of Sweden establish the "prize in economic sciences in memory of Alfred Nobel." Following lengthy debates about this prize, which many do not consider to be "a real Nobel Prize" because it was not originally mentioned in the founder's will, the Foundation decided that no further prizes could be added in the future.

The number of subjects was limited, as were the number of prize winners: up to three per year in each subject, with a total of up to nine in the sciences per year. There is also the possibility—which has occurred six times in physics, eight in chemistry, and nine in medicine—of the prize not being awarded to anyone or of it being kept aside for the following year.

This limit to certain subjects and to an extremely limited number of recipients per year fueled the perception of the exclusive and thus prestigious nature of the prize.[2]

So one must study physics, chemistry, or physiology or medicine. In the next chapter, we will discuss the manner and flexibility with which—at different times in history—these subject boundaries were interpreted. For now, if one is interested in winning the Nobel Prize, these are the subjects to choose.

Another fundamental condition is necessary in order to win the prize: you must be alive! The prize cannot be awarded posthumously, at least not now. In 1974, the Foundation changed its statutes in order to exclude the possibility of awarding a prize to the deceased. Before then, two times—once in 1931 for literature (Erik Axel Karlfeldt) and once in 1961 for peace (Dag Hammarskjöld)—the candidate had passed away before the official announcement.

Technically, there are no restrictions on the gender, age, or nationality of recipients. However, one has only to take a look at the honor roll of laureates to note that very few women have received a Nobel. Of the 965 individuals who were awarded prizes between 1901 and 2023, only sixty-five women received the prize, and of these only twenty-six in sciences (out of the 621 laureates).[3]

And what about age? There are no established limits, however the average age of the Nobel laureates is fifty-nine years, and the majority are awarded the prize between the ages of sixty and sixty-four.[4]

As for nationality, Nobel was extremely clear in his will:

it is my expressed desire that the prizes be assigned without taking into account the nationality of the candidates, in order for them to be awarded to the most deserving, be they Scandinavian or not.

However, it is immediately clear that some countries get the larger share of prizes. Obviously, it depends on which country we consider a scientist's own, since many scientists have worked in different countries throughout their careers. In accordance with Nobel's wishes, the Foundation officially indicates only the birth country of the laureates. From 1901 to 2020, there have been 281 laureates in total (197 for sciences) born in the United States, 89 in the United Kingdom (68 for sciences), and 78 in Germany (53 for sciences, to which one should add others who were born in then-German territories). If one looks at candidates' places of residence at the time of the prize, the substance does not change much, at least for the countries with the highest number of laureates: the United States, the United Kingdom, and Germany are again the most highly represented in all three science domains subjects.[5]

In summary: three of every ten Nobel Prizes for sciences were awarded to residents of the United States. And out of every two prizes awarded, one went to a resident of the United States, the United Kingdom, or Germany. From 1935, at least one Nobel Prize was awarded to residents of the United States every year. The data on the prizes can be seen as a significant indicator of the capacity to attract the best researchers on an international scale. This is especially true for the United States. For every three laureates who are US residents, one was born elsewhere. In 2016, six "American" Nobel prize recipients were born in other countries.

The most promising places to study or work depend on the subject and also on the specific point in time. Certain German universities, Cambridge University in the United Kingdom, and the Pasteur Institute and Sorbonne in Paris each had their heyday as incubators of Nobel Prizes in certain subjects. After the Second World War, they were largely surpassed by a few American universities: the California Institute of Technology, the Massachusetts Institute of Technology, and Harvard University.

And yet above all of these advantages, the greatest one is becoming the student of a Nobel. In her extensive study of ninety-two US Nobel prizes in the sciences, the sociologist Harriet Zuckerman showed that forty-eight had been students or collaborators of another Nobel Prize.[6]

The most striking example is New Zealand physicist Ernest Rutherford's "Nobel school." Eleven of his "guys" (as he called his assistants) have gone on to receive a Nobel, among them Niels Bohr, Patrick Blackett, James Chadwick, John Cockcroft, Piotr Kapitsa, and Ernest Walton. A twelfth, Henry Moseley, was very close to winning the prize before he tragically died (see chapter 3).

I will briefly note that some historians of the Nobel Prize and its recipients have classified the latter by religion, noting that a significant proportion of Nobel laureates have been of the Jewish and Protestant religions. During the first century of the prize, at least a fifth of the prizes were awarded to Jewish scientists. In fact, 116 have been awarded prizes—thirty-six for physics, twenty-two for chemistry, and thirty-nine for medicine.[7]

To win a Nobel Prize, one has to study, work, and aim for results that are innovative and recognized by one's peers. According to Eugene Garfield, one of the pioneers of bibliometry and scientometrics and therefore one of the first people to develop quantitative indicators for research activity, future Nobel Prizes may reliably be found among the 1 percent to 2 percent of the most cited scientists in each sector. "Almost without exception, Nobel Prize laureates have total citation counts approximately fifty times higher than the average scientist."[8]

On the other hand, there are examples of widely cited scientists and works that have never received the Nobel Prize. Oliver Lowry, an American biochemist and author of the most cited scientific article of all time, was never nominated for the Nobel Prize. This is true for many authors of the hundred most-cited scientific articles in history.[9]

There are also exceptions as far as specific publications are concerned, with findings that were later recognized in Stockholm initially struggling to reach publication. In 1937, the journal *Nature* said there was not room to publish the letter in which Hans Krebs presented his discovery of the citric acid cycle (known today as the "Krebs cycle"), which was awarded a prize for medicine in 1953. The same journal did not publish Kary Mullis's manuscript on the polymerase chain reaction (PCR) technique destined to revolutionize molecular biology, which was awarded the chemistry prize in 1993.

In many cases, other prizes and scientific recognition preceded the awarding of the Nobel Prize. During the first fifteen years of the Nobel, all the French laureates had already received prizes and medals from the Académie des sciences and only two British scientists (William and Lawrence Bragg) had not already received one of the three main medals from the Royal Society (the Copley, the Rumford, and the Davy medals). During the two years preceding her winning the Nobel Prize for medicine (1983), the American biologist Barbara McClintock received more prizes than she had ever received throughout her entire career.[10]

There are, however, some examples of scientists who had to wait until they received the Nobel Prize before getting full recognition from their colleagues.

In any case, if one looks at just the statistics, the identikit of the typical Nobel laureate seems pretty straightforward: a middle-aged male scientist, active in one of the most prestigious universities in the United States (or a handful of other countries), often a former student or a collaborator of another Nobel laureate, author of influential and original articles, and often a previous recipient of significant prizes and recognitions.

Do you have none of these characteristics? And more to the point, do you have no chance of going to one of the universities where Nobel Prizes flourish or of working side by side with another Nobel laureate? Do not lose hope, and read the following pages.

HOW TO WIN A NOBEL PRIZE BY NOT TAKING HOLIDAYS OR PUBLISHING A SINGLE LINE

It is the summer of 1958. Jack Kilby, an engineer from Missouri, is employed by Texas Instruments. Ten years earlier, Bell Laboratories introduced the transistor, which calculators required in large numbers. When experts looked at the future of technology, they imagined it as a science fiction film: full of increasingly large, complex machines and calculators. Companies like Texas Instruments had hangars packed with workers, most of whom were women, sealing small pieces of silicon under microscopes. The wires often broke or disconnected, and it took only one mistake for the whole process to require starting over. The more complicated the circuits, the higher the probability of mistakes.

As a new employee, Kilby has no holiday time and stays in the office while most of his colleagues go on vacation. Perhaps the temporary absence of workers gave him the idea for an alternative to the laborious process, or maybe the silent office, empty of the usual tasks and distractions, allowed him to work and reflect calmly.

On September 12, 1958, he calls on his suntanned and rested colleagues to give them a demonstration. In his hand, he holds a small 5×18 cm object made of germanium, aluminum, and gold. To avoid assembling separate elements, Kilby has built resistors, transistors, and capacitors on a single semiconductive support: it is the birth of the so-called integrated circuit. Six months later, another engineer, Robert Noyce, duplicates Kilby's result and adapts it to larger-scale production. Silicon, less expensive and more widespread, would soon replace the germanium originally used by Kilby (which is why we do not call it Germanium Valley nowadays). This was the first step toward microchips and the era of microelectronics in which we now live.

The industry is skeptical at first. The military sector is the first to see the potential of this innovation. The US Air Force uses Kilby's integrated circuit in its first computer in 1961.

In the mid-1960s, the classic office calculator is still the size of a typewriter and costs approximately $1,200, half the price of a family car. Kilby, who by then is a well-established figure in the company, secretly accepts his boss's challenge: to create a "calculator that can fit into a coat pocket" that costs less than $100. The "Pocketronic" entered the market in 1971, costing $150. That year, over five million are sold. In following years, the price continues to decrease, as does its size. Sales double every year.

Kilby remained relatively unknown to the public, but few innovations have changed the contemporary world as much as his. His idea was not to add but to simplify, not to assemble but to integrate. His intuition overturned the technology and imagination of those times and launched our world toward miniaturization and portability.

For "his role in the invention of the integrated circuit," Kilby was awarded the Nobel Prize for physics in 2000 at the age of seventy-seven, five years before his death.

His story contradicts most of the "typical" characteristics of a Nobel laureate, except that he was an American man. Kilby was not an academic (he did not even have a PhD), nor did he work in a prestigious university. He was named "distinguished professor" at Texas A&M University only after he became famous for his invention. He had never published a scientific article. The Academy of Sciences decided to reward an invention made in industry as opposed to a scientific result or publication, which is rare but not unique in the history of the prize. Without his invention, there would be no Apple or Google today.

NEVER JUDGE A NOBEL BY APPEARANCES: THE COUNTRY DOCTOR AND THE "MOM FROM SAN DIEGO" WHO WON THE NOBEL PRIZE

Toronto, Canada, 1920. Frederick Banting is a thirty-nine-year-old country doctor with no research experience. He has barely completed his university studies because of a war injury to his arm caused by a piece of shrapnel. For years, however, he has been studying an illness for which there is no treatment: diabetes. The illness is known to be linked to pancreatic metabolism, and Banting was struck by reading about a clinical case in a journal: in a patient who was affected by a near total atrophy of the pancreas but in whom the so-called Islets of Langerhans had remained intact, there was no insurgence of diabetes. Banting is not the first one to speculate that the regulation of sugar levels depends on the internal secretion of the islets; but until then, all trials to use pancreatic extract for therapeutic means had failed, probably because the extract was destroyed by the enzymes produced in the remaining pancreas. At 2 o'clock in the morning of October 3, 1920, Banting writes in his notebook: "Tie up the pancreatic duct of a dog. Wait six to eight weeks. Prepare the extracts."

Banting obtains a meeting with John Macleod, a professor at Toronto University. The professor is quite skeptical, but Banting convinces him to put a small laboratory at his disposal. Macleod concedes eight weeks and ten animals. This is not enough, but taking advantage of the fact that Macleod is away for a few months, Banting goes ahead and uses all of his personal resources. He leaves his job as a doctor and sells his car to acquire new animals. He even finds a graduate, Charles Best, who is willing to help him. The two pursue the experiments while living on Banting's severance package from the army.

They see their first successes during the summer of 1921. Banting and Best extract a substance from the pancreas that they call "insulin"; they inject it into a diabetic dog whose symptoms improve. At that point, Macleod becomes interested and grants Banting additional resources to test his theory on humans. First, Banting and Best inject themselves with insulin: other than a feeling of slight weakness, the two do not suffer from any serious side effects. In January 1922, they try it on a fourteen-year-old boy on the verge of death because of his illness: the patient recovers quickly. The therapy gives excellent results in other patients, and the extract, patented by the researchers, starts being produced on a wide scale.

A year later, in 1923, Banting and Macleod are awarded the Nobel Prize for medicine "for the discovery of insulin." Banting is furious that Macleod, who he felt had not contributed to the discovery, is also awarded the prize and threatens to refuse it up to the last minute. He finally accepts the Nobel Prize but shares the money with his colleague Best, with whom he has done most of the research. Macleod does the same with another member of the research group, Bertram Collip.

Barely three years after his nocturnal notes in his notebook, Banting went from being an unknown country doctor to a Nobel laureate. His intuition and his tenacity led to an invention that to this day has benefited many patients.

Unlike Banting, Maria Goeppert undertakes a formal career in university research early. Her journey, however, turns out to be a rather bumpy one. She was born in 1906 in Kattowitz, Germany (now Katowice, Poland). The school she attends in Göttingen is forced to close, victim of the grave German financial crisis. In 1930, Goeppert is awarded a doctorate in theoretical physics, but in those years, it was impossible to obtain a university position, especially for a woman.

After graduating, Goeppert marries an American chemist, Joseph Mayer. She follows him to Baltimore, where he has university work, and then to New York and Chicago. In the years following the Great Depression, finding even one university position is a challenge, never mind two. Goeppert's presence in the department is tolerated at best, especially in the beginning. She develops an interest in chemical physics and continues with her own research, undertaking minor tasks with no formal recognition. In Chicago, she is finally given the title of professor and with that obtains her colleagues' respect (among them was Enrico Fermi) but still no salary. Starting in 1948, Goeppert-Mayer develops a theory explaining how some chemical elements have more stable nucleuses than others. Her model identifies a series of shell-like structures in the nucleus, similar to the layers of an onion. Depending on the filling of the different shells and, in particular, the configuration whereby the number of protons or neutrons equal one of the so-called magic numbers (2, 8, 20, 28, 50, and 82), the nucleus is more or less stable. Following this important discovery, Goeppert-Mayer finally obtains her deserved recognition: a university post at the University of San Diego and finally the Nobel Prize for physics in 1963, divided between her and Hans Jensen, who independently reached the same conclusions. The two are awarded a quarter of the prize each. The other half of the prize is awarded to Eugene Wigner "for his contributions to the theory of the atomic nucleus and the elementary particles, particularly through the discovery and application of fundamental symmetry principles."

Goeppert-Mayer is currently one of only five women to have been awarded the Nobel Prize for physics (the others are Marie Curie in 1903 and 1911, Donna Strickland in 2018, Andrea Mia Ghez in 2020, and Anne L'Huillier in 2023). Not bad for a scientist who for many years was unemployed and forced to undertake her research as a personal hobby. In announcing the news, one local newspaper's headline read "San Diego Mom wins Nobel Prize."

IT IS NEVER TOO LATE (OR TOO EARLY) TO WIN A NOBEL PRIZE

So have the stories of Kilby Banting and Goeppert reassured you?

Let us try to find some other reasons for optimism. Almost half of the prizes to women have been awarded during the past fifteen years. If this trend continues, the chances for women to win a prize should increase.

If age is worrying you, either because you feel too old or too young, fear not: it is never too late (or too early) to win a Nobel Prize.

We already mentioned Kilby, who was awarded the prize for physics at the age of seventy-seven, more than forty years after his invention. In 2002, Raymond Davis Jr. received the prize for physics at the tender age of eighty, "for pioneering contributions to astrophysics, in particular for the detection of cosmic neutrinos." The oldest winner of the prize for chemistry was the eighty-five-year-old John B. Fenn, who won the prize in 2002 "for the development of methods of identification and structure analyses of biological macromolecules." In medicine, Peyton Rous was awarded the prize at the age of eighty-seven in 1966 "for his discovery of tumour-inducing viruses." The oldest winner of any Nobel in the sciences is Arthur Ashkin, awarded the prize in physics in 2018 at the age of ninety-six for "groundbreaking inventions in the field of laser physics."

And how about the younger winners? Until now, only one scientist has succeeded in winning the prize before the age of thirty. In 1915, William Lawrence Bragg won the prize at the young age of twenty-five with his father Sir William Henry Bragg "for their services in the analysis of crystals by means of X rays." Before you judge this a case of nepotism, wait until you hear more about this young man.

Lawrence Bragg was probably a predestined child. At the age of five, he fell off his tricycle and broke his arm. His father, inspired by the recent Röntgen discovery, was conducting experiments on X-rays in his laboratory and used them to analyze the fracture.

A precocious talent, his role in the Nobel discovery was long thought to be minor, as if his father's experience had led the way for his recognition. But in reality, his was the more significant of the two contributions, especially to the theory of physics. He was the one, based on the experiments by German physicist Max von Laue (Nobel Prize for physics in 1914), who refuted the corpuscular theory of X-rays supported by his father in favor of a wavelike explanation. He had an intuition while walking along the river in Cambridge: "Laue's spots were due to the reflection of X-ray pulses by sheets of atoms in the crystal."[11] He formulated what is now known as Bragg's law, which established a connection between wavelength, distance between lattice planes, and diffraction angle. His father, on the other hand, played a crucial part in the development of sophisticated experimental instruments.

"My son," William proudly wrote in *Nature* in 1912, "has given a theory which makes it possible to calculate the positions of the spots for all dispositions of the crystal positions and photographic."[12] However, in Brussels in 1913, at one of the prestigious Solvay conferences on the structure of matter dedicated to the discoveries by Bragg Jr. and Laue, only his father was invited: Lawrence, who stayed at home, had to make do with a postcard congratulating him, signed among others by Marie Curie and Albert Einstein. The 1915 prize ceremony was postponed due to the war; however, Lawrence would have not been able to attend since he was commissioned in the Royal Horse Artillery. This year marked one of the most tragic and controversial moments for the Nobel Prize (see chapter 3).

Other scientists, especially physicists, have been awarded the prize at the beginning of their careers. Werner Heisenberg (1932) and Paul Dirac (1933) were both thirty-one at the time of their awards. Other precocious laureates were thirty-four-year-old James Watson (medicine in 1962) and thirty-five-year-old Guglielmo Marconi (physics in 1909). Among those who won a Nobel Prize before the age of forty were Marie Curie, Ernest Rutherford, Niels Bohr, and Enrico Fermi. Frederick Sanger was an almost unique case, having received the prize for chemistry in 1958 at the age of forty: this relatively early recognition "gave him the time," as it were, to win another Nobel Prize, also for chemistry, in 1980.

Incidentally, if you believe that fate is sealed on the day we are born, you should know that the most common dates of birth among the Nobel laureates are May 21 and February 28.

THE LONG JOURNEY TO STOCKHOLM

Let us suppose that at this point you have already made a significant discovery. You have published your article in one of the most prestigious international scientific journals, and you justifiably are feeling satisfied. Well, your potential journey to Stockholm has just begun. Make yourself comfortable at least until October. This is the month during which the Nobel machine is set in motion. Now you must hope that a colleague who respects you or who has read your article will be one of the three thousand people to receive the much-coveted letter under the letterhead of the Royal Swedish Academy of Sciences or of the Karolinska Institute inviting them to nominate candidates for the Nobel Prize. Who receives this letter? Former Nobel laureates first of all. For the physics prize, for example, potential nominators include:

1. Swedish and foreign members of the Royal Swedish Academy of Sciences;
2. Members of the Nobel Committee for Physics;
3. Tenured professors in the Physical sciences at the universities and institutes of technology of Sweden, Denmark, Finland, Iceland and Norway, and Karolinska Institutet, Stockholm;
4. Holders of corresponding chairs in at least six universities or university colleges (normally, hundreds of universities) selected by the Academy of Sciences [. . .]; and
5. Other scientists from whom the Academy may see fit to invite proposals.[13]

The deadline to suggest a candidate is January 31. If your name is not submitted or is submitted after that date, you will have to wait until the following year. If someone who has not been invited to nominate puts your candidacy forward, it will be ignored. If you receive the letter and were thinking of nominating yourself, you cannot, since self-nominations are not permitted.

In February, the days in Stockholm are getting longer. Your name has finally been put forward! But the road ahead is still uphill, and approximately five hundred of your eminent colleagues have also been nominated. Now the ball is in the court of the five members of the respective Nobel committees (chemistry, physics, and physiology or medicine). They are the ones who must draft an initial list of potential candidates. They will solicit the opinions of experts who also work in the subfields of the most promising candidates. During the summer, while you might have been on holiday, the members of the Nobel committees start writing reports on each candidate as well as a general report.

It is now the end of September, and almost a year has gone by. Now is the time for the final recommendations. The committees put forward the names of their nominees in the different subjects to the approximately four hundred colleagues of the Academy or the approximately fifty colleagues of the Nobel Assembly of the Karolinska. At the beginning of October, there is a plenary vote. All members of the Academy of Sciences, for example, vote on the prize for physics or chemistry, regardless of their own area of expertise. It can happen, and has, that a majority vote goes against the committee's specific recommendation; but in practice, it occurs rarely, especially in recent decades. The recipient's names are usually announced on the day of the vote to avoid any leaks.

If you receive an incoming call with the Stockholm area code, congratulations: you have made it! You will now have to give a long series of interviews and meetings with the press. But your main goal during the next two months, other than answering (or avoiding) questions from the media, is to stay in good health. Some laureates have not been able to attend the ceremony due to health reasons or, worse, because they passed away before the official award ceremony. If that befalls you, your name will still be included in the annals of the prize.

At the beginning of October 2011, the Nobel Prize for medicine was announced for Ralph M. Steinman "for his discovery of the dendritic cell and its role in adaptive immunity." The difficulty of contacting him to give him the good news was soon explained: Steinman had passed away three days earlier. Based on the Statutes of the Nobel Foundation whereby no prize could be awarded to deceased scientists unless the award had been announced beforehand, it was decided that the prize to Steinman would be maintained since the Karolinska had announced it before they discovered that he had passed away.

But let's get back to you: during the rest of the phone call, which you will probably not be listening to very carefully, you will be asked to keep your schedule free during the week of December 10.

AN UNFORGETTABLE WEEK

December has arrived. A busy week of engagements awaits you in Stockholm.

On your arrival, generally December 6, a driver and a Swedish diplomat who will be at your disposal all week will welcome you. December 7 is dedicated to press conferences and various receptions. On December 8, you will be expected to undertake your sole obligation established by the statutes of the prize: to hold a conference at the University of Stockholm, which is not an unusual undertaking for a scientist. You can talk about the discovery for which you have received the Nobel Prize or about another part of your research. The physicist Ernest Rutherford, with his usual love for a *coup de théâtre*, took advantage of the occasion to announce a brand-new discovery.

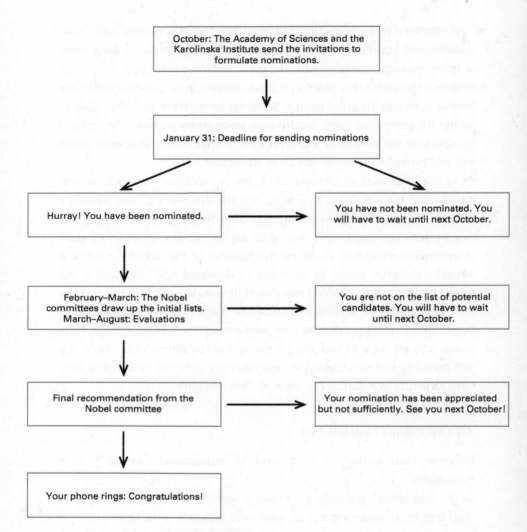

2.1 Summary: How to win a Nobel Prize.

The other engagements that await you are definitely more peculiar, to say the least. But the instructions and the trials are meticulous.

The awards ceremony in which you will have a leading role takes place on December 10 at the Konserthuset (until 1925, the ceremony took place at the Royal Academy of Music).

Following a speech by the representative of the Nobel Foundation and a series of music pieces, a member of the Nobel committee of the

Academy of Sciences (or of the Karolinska) will introduce you and explain the discovery or the result for which you have been awarded the prize.

When you hear your name, you must stand up, walk up to the stage, and stop at a precise spot. Here you receive a diploma and a medal from the hands of His Majesty the King of Sweden, Carl XVI Gustav. The diploma includes your name and the "quote," or the reason for which you have been awarded the prize. Each diploma is designed by a Scandinavian artist and calligraphist. The medal is made of recycled 18 karat gold (it was made of 23 karat gold until 1980), weighs 175 grams, and has a diameter of 66 millimeters. On one side, it shows the portrait of Alfred Nobel and his birth and death dates. On the other, for the prizes in physics and chemistry, nature is represented by a goddess emerging from the clouds, holding a cornucopia while the "genius of sciences" reveals her face by lifting a veil. On the medal for medicine, there is a portrait of the "genius of medicine" with an open book on his lap, concentrated on capturing water flowing from a rock to quench the thirst of a young girl. All prizes bear an inscription from the sixth canto of the *Aeneid*: "Inventas vitam iuvat excoluisse per artes" (loosely translated as "They who bettered life on earth by their newly found mastery") and carry the signature of the respective academies or institutions. The king also hands you a document that confirms the financial value of the prize, which amounts to approximately eleven million kronor nowadays (just over 951,000 euros at the exchange rate in late 2024).

The amount for the prize is linked to the return of the financial investments made with Nobel's original legacy. Today it is on par with the original amount. At times, during the First World War and again during the Second World War and the 1980s, it fell significantly, to a little more than two million kronor. The amount peaked during the first years of this century, when the prize winners received almost twelve million kronor. Therefore, you could have done better but also much worse.

The sum you receive also depends on whether you are the only recipient for your category or if you have to share the prize with one or two other colleagues. Keep in mind that since there can be a maximum of three scientists who can be awarded the prize for one subject, you could also receive a quarter of the prize according to the bizarre Nobel arithmetic.

The Academy can award half the prize to one individual and the other to two individuals (generally because they have worked together), thus dividing the second half of the prize into two parts.

The award money does not have to be used for research; you are allowed to do what you wish with it. Initially, the award so dwarfed a scientist's customary budget that many invested the money in their research (like Marie Curie) or donated it to their institution—as did Röntgen, one of the very first Nobel laureates, who gave the money to the University of Wurzburg. More recently, laureates have used the money to fund other prizes or scholarships for sciences, as did Paul Greengard (Nobel Prize for medicine in 2000), Christiane Nüsslein-Volhard (medicine in 1995), and George Smoot (physics in 2006). Others donated the money to charity organizations not focused on science: Günter Blobel (Nobel Prize for medicine in 1999) used it to have a cathedral restored and a synagogue built in Dresden. And sometimes award recipients give more back to the Nobel Foundation than they receive. The Nobel for medicine Georg von Békésy named the Nobel Foundation itself as the heir of his personal art collection, which was worth ten times the amount of money he had received.

Most recipients have used their sum of money for personal expenses and investments: homes, children's education (like Wolfgang Ketterle, Nobel for physics in 2001), a powerful motorbike (Paul Nurse, the medicine prize in 2001). Richard Roberts (Nobel for medicine in 1993) even spent it on the creation of a croquet lawn in his back garden! Serge Haroche, the physics prize in 2012, admitted three years later to having been so busy after receiving the prize that he still had not had time to think about how to spend the money. Another physicist, Richard Feynman (Nobel in 1965) announced that he wished to save the money in order to pay his taxes for the rest of his life.

Speaking of taxes, do you want to know if the prize will be taxed? It probably will, though it depends on where you live. Since 1986, the prize for Americans is taxed like any other form of revenue. But consider yourselves lucky: in 1923, the Austrian Inland Revenue took two-thirds of the prize destined to the Nobel for chemistry Fritz Pregl. In 1922, Albert Einstein made the entire sum available to his former wife, Mileva Marić, per their agreement at the time of their separation.

How much is the medal worth? Based on the quantity of gold it contains, its value is estimated at $5,500. But obviously, its symbolic value can hugely increase the sale value. In 2015, the medals belonging to Francis Peyton Rous (Nobel for medicine in 1966) and Alan Lloyd Hodgkin (Nobel for medicine in 1963) were sold at two auctions in the United States, respectively, for $461,000 and $795,000. The sales of the 1962 Nobel Prizes for medicine belonging to Francis Crick and James Watson were even more astonishing. Crick's medal was sold by his heirs in 2013, nine years after his death, for $2.3 million; the Chinese biotech investor Jack Wang bought it. Watson's medal sold at auction for the record price of $4.7 million in 2014. The scientist declared that he had sold it for financial reasons after he was excluded from the boards of numerous companies and his requests for public conferences collapsed, following comments he made on the relationship between IQ and race. Alisher Usmanov, the Russian billionaire who bought it, however, decided to return the medal to him after he acquired it.[14]

During the Nazi regime, the Nobel Prize winners for physics, Max von Laue (1914) and James Franck (1925), sent their medals to Niels Bohr in Copenhagen for safekeeping. When the German soldiers came to search Bohr's institute in 1940, they were unable to find them. They were actually right under their eyes, transformed into an orange liquid. The Hungarian chemist George de Hevesy, who had sought refuge in Copenhagen because of his Jewish origins, had melted them in *aqua regia*, a mix of nitrite and hydrochloric acid. After the war, Hevesy (Nobel prize for chemistry in 1943, although not for this undertaking) managed to recover the gold, and the Academy of Sciences used it to remake the two medals.

But surely you can't be worrying about money while you stand just a few inches from the king of Sweden? At this point, you must shake his hand, bow to his majesty, bow to the laureates from the previous years, and finally bow to the rest of the public.

Now, together with the other approximately 1,300 other illustrious guests (of whom you were allowed to invite about fifteen), you will make your way into the so-called Blue Room of the Stadhuset, the Stockholm Town Hall, for the banquet. The evening is shown live on Swedish state television. Every year, about 250 places are reserved for students, who can take part in a lottery that allows the winners to buy a ticket for the banquet for approximately 600 euros. The room (which is not blue; the architect had second thoughts about the color at the last minute) is decorated with more than twenty thousand flowers from Sanremo, the town where Nobel died. The dress code is very formal, black tie for men and evening gowns for women.

The evening's menu is kept secret until the last minute. It generally consists of three courses, and the dessert is often (but not always) a so-called Glace Nobel, on which the letter N appears in memory of the founder. When coffee is served, the laureate or one of the laureates (generally the eldest) of each field makes a brief thank-you speech. A toast is raised to His Majesty and another to the founder, Alfred. At the end of the banquet, you will go through to the Golden Room for the ball, where the famous photograph of Francis Crick dancing with his daughter Gabrielle was taken in 1962.

The next day, you will receive your check. In the evening, there is another banquet at the Royal Palace in honor of the Nobel laureates. On this occasion, four courses are served, and they generally include a roebuck shot by the king himself during a traditional autumn shoot.

Did you sleep well, or did you have a glass too many during the various banquets? Do not panic if, at 6 a.m., someone opens the door to your room at the Grand Hotel opposite the Royal Palace. And don't be surprised if a small procession led by a female figure followed by pageboys and maids of honor and children dressed as elves crosses your room holding candles and singing in front of your bed to offer you Christmas cakes and coffee. This is a small surprise organized by the hotel on the day of Santa Lucia, though it will not come entirely as a surprise. Since one of the laureates almost had a heart attack, possibly exhausted by all the previous emotions, the Grand Hotel now asks all guests if they wish to be woken up this way or not. Even Nobel Prizes are allowed to rest at times.

3

A PRIZE WITH NO BORDERS?
NO POLITICS PLEASE, WE ARE SWEDISH (SCIENTISTS)

The atom was used for war,
For victory over the planet.
You wonder: whose fault is it?
The answer is: Otto Hahn's.
The Academy chose the best one.
A German is the winner.
You wonder: whose fault is it?
The answer is: Otto Hahn's.
—Song by the German physicist prisoners at Farm Hall, 1945

A PRIZE WITH NO BORDERS—OR SCIENCE AS A PROSECUTION OF POLITICS BY OTHER MEANS

In his brief will, Alfred Nobel made clear that the prizes were to be assigned without any consideration of the nationality of the candidates.

Thus it was not, as some Swedish academics had hoped and even claimed, a prize or financial award reserved for the scientists and institutions of his birth country.

It was a prize with no borders, destined to represent the purely intellectual nature of scientific research and its capacity to overcome national divisions. From this point of view, the fact that it was adjudicated in Sweden was ideal. Even before Sweden emerged as a politically neutral country after the First World War, the perceived impartiality of the prize was reinforced by the fact that the deliberating institutions were not linked to any of the three spheres of cultural and scientific influence at the time: German, French, and Anglo-American. In 1911, the Swedish scientist Allvar Gullstrand opened his speech at the Nobel banquet by explicitly comparing the prize to the Swedish flag.[1]

Ironically, the media and the public immediately interpreted the prize in light of international competition. Countries with Nobel Prize–winning citizens trumpeted their achievements. Thus,

the prevailing climate of nationalism [. . .], achievements that supposedly had benefitted all mankind were heralded not just as those of individuals belonging to particular nations [. . .] but as proof of the superiority of that nation in the realm of culture or science.[2]

While commenting on the first year of the prize and the absence of British laureates, a French newspaper wrote that, notwithstanding,

The British have the insolence to pretend that they are, in all respects, the first nation on this earth. . . . But England has been defeated by the rest of Europe in the peaceful pursuits of the sciences and the arts—that is, in the domains which represent true civilizations and real progress.[3]

Within days, English newspapers like *The Times* were flooded with letters that complained about the exclusion of their countrymen from the first prizes.[4]

When in 1905 all of the scientific prizes went to German scientists (Philipp Eduard Anton von Lenard for physics, Robert Koch for medicine, Adolf von Baeyer for chemistry), several German newspapers used the nationalistic headline "Germany Ahead in the World!"[5]

During the first ten years of the prize, a German paper began compiling statistics on the laureates based on their nationality. This has now become a common feature in most of our media whenever the prizes are announced or referred to. Germany was well ahead in the list, and the newspaper wished for every effort to be made "to ensure that in the future Germany would maintain the brilliant position obtained during the first decade." The same article pointedly noted the poor showing by American scientists. Only two received the prize during the first ten years, one of whom was born in Prussia (Albert Abraham Michelson, Nobel Prize for physics in 1907): "A nation of ninety-three million that receives only one thirtieth of the prizes!"[6]

The patriotism elicited by the Nobel Prizes also led to expressions of shame for having neglected leading scientists at home. For example, the French newspaper *La Liberté* described the Nobel Prize awarded for physics to the Curies as "a good lesson to those among us who exploit favors and official positions" and in general to a country that "does not know its scientists."[7]

The media continued to provide many examples of the national pride with which the prizes are received. "Americans from Purdue and Harvard Win—German and Pakistani Cited" was the strange headline in the *New York Times* at the time of the Nobel Prize announcement of 1979.[8]

The status of Nobel Prize winners as "national heroes" garnered them commemorative stamps of individual laureates and groups from the same country. The attention paid to Nobel prize recipients exemplifies the ability of intellectual heroes to incarnate national unity and identity identified by Thomas Carlyle in his classic study on heroes. For Carlyle, authors like Shakespeare or Dante gave their nations an "articulate voice": "voice of genius, to be heard of all men and times [. . .] King, whom no time or chance, Parliament or combination of Parliaments, can dethrone" and "that can keep all these together into virtually one Nation" and allow us to say "we are of one blood and kind with him"[9]

In the prize's more recent years, countries aiming to improve or reinforce their position in the sphere of scientific research publicized their desire to have more nationals earn Nobel Prizes.

In 2004, at the peak of national enthusiasm for research in stem cells and animal cloning announced by Professor Woo-suk Hwang, the Korean government created a committee with the explicit objective of increasing the chances of receiving a Stockholm prize for their scientists in order to ensure a "Nobel Prize for the Nation."[10]

In 2001, the Japanese government set itself the goal of thirty Nobel Prizes within fifty years—and completed half that goal during the first fifteen years, although, according to many commentators, the result was largely independent of government effort, since most prizes were for results obtained from research begun in the preceding decades.

In 1998, some members of the Chinese Academy of Sciences launched an appeal—"Chinese science: the time has come for a Nobel Prize!"—that was immediately echoed by an invitation from the director of the National Natural Science Foundation of China "to make a plan in order to obtain a Chinese Nobel Prize at the beginning of the twenty-first century." Similar announcements marked the beginning of a real "Nobelmania" characterized both by research policy initiatives and public anticipation. A public opinion poll in 2000 revealed that the awarding of a Nobel Prize to a scientist was considered "one of the most probable events in China in the next ten years."

Many academics have argued that this boost of enthusiasm in China for the Nobel should be interpreted from a political point of view. "To win a 'homemade' Nobel Prize is a way of saving face for the Chinese political leadership,"[11] a way of proving that Chinese scientists do not need to leave the country in order to achieve significant results, as did Tsung Dao Lee and Chen Ning Yang (Nobel Prizes for physics in 1957), Daniel C. Tsui (Nobel Prize for physics in 1998), and Charles Kuen Kao (Nobel Prize for physics in 2009).

The "homemade" prize finally arrived in 2015. "Homemade" is an apt description, although it is difficult to ascribe the prize to strategies and investments by the Chinese government. The prize for medicine was awarded to an eighty-five-year-old researcher, Tu Youyou, who had no doctorate and an unconventional, nonacademic career path. Born in 1930 in Ningbo, on the eastern coast of China, at the age of sixteen Tu Youyou was affected by a serious form of tuberculosis. Once she was cured, she decided to dedicate herself to medicine. In 1969, in the midst of the Cultural Revolution, Tu Youyou was recruited for a project known in code as project "523." The goal was to address the diffusion of malaria that was devastating North Vietnamese troops supported by China. Tu Youyou found the answer to the problem by going back with contemporary pharmacological methods to traditional Chinese medicine, discovering that the *Artemisia annua* plant actually had antimalarial properties as described in ancient Chinese medical texts. As the award announcement noted, "Artemisinin-based medicines have led to the survival and improved health of millions of people."

The Chinese "Nobel obsession" can be interpreted as

part of China's resurgent nationalism, as with winning the right to host the Olympics. [. . .] The nation is more willing than ever to embrace both the Olympics and the Nobel Prize, and for much the same reason. It is widely believed that, until now, China has not been given these symbolic awards, because the international community has not fully acknowledged China's place in the world.[12]

THE OLYMPIC GAMES OF SCIENCE?

So the Olympics. In the same years during which Alfred Nobel is writing his will, the twenty-nine-year-old secretary of the Union of French Athletic Sports Societies, Baron Pierre de Coubertin, undertakes a project that at the time seems insane rather than visionary: he hopes to bring the Olympic Games back to life per the tradition of classical antiquity. On June 16, 1894, Coubertin succeeds in convening representatives of twelve states in the halls of the Sorbonne. In his invitation letter, he presents his project as

the happy consecration of the international understanding that we are striving for, or failing its creation, at least preparing for. The reestablishment of the Olympic Games, on bases and conditions that are in line with the necessities of our modern life, would put representatives of the various world Nations in competition against one another every four years, and it is possible to believe that these pacific and courteous competitions would help build the best possible internationalism.[13]

The ambitious project risks collapse almost immediately, when the Union of Gymnastic Societies of France threatens to leave if the German delegation—a historic political enemy—attends. Once this problem is solved (Germany sends only an observer) on June 23, the seven points of the "Olympic Chart" are unanimously approved. Just under two years later, on April 5, 1896, the same year Alfred Nobel dies, the first modern Olympics are inaugurated.

Thus, another extraordinary invention took place during the years that Alfred Nobel's prize was created and shared the same international spirit and idealistic elements.

As science historian Sven Widmalm notes, "Nobel's belief in progress through pacific international competition" was perfectly timed with the great universal exhibitions and the relaunching of the Olympics.[14] Both the Olympics and the competition for Nobels were opportunities for nations to have proxy competitions against each other, more or less self-consciously, as an "antidote to militarism." In those years, science, in particular, was starting to be considered from a "political angle, comparing and evaluating the contributions from different nations in terms of prominence and inferiority, domination and dependence."[15] It is an inheritance we still see today, when we list the Nobel Prizes by national citizenship or rank universities and countries by scientific excellence (see chapter 1). As early as 1931, the *New York Times* published a list of Nobel Prizes per country under the headline "Germans Lead the Nobel Prize Winners; Americans Now Stand Fourth in the List, Following the French and British."

The Nobel Prize's capacity to incarnate this "political rivalry by other means" was reinforced by the accreditation of certain institutions and scientists as impartial arbiters of the competition, thanks to Sweden's neutral political status. Thus, the "members of the Royal Academy of Sciences and of the Karolinska became perpetual juries of a perpetual universal exhibition."[16]

The Nobel Prizes have even been explicitly compared to the modern Olympics, not only in the popular press, which has called the Nobels the "science Olympics" or the "Olympic Games of the mind."[17] In 2012, the Chairman of the Nobel Foundation, Marcus Storch, opening the 2012 awards ceremony, described the Nobel awards as "the first international prize, an era in which the Olympic Games had been re-launched following a pause of 1500 years" and described the Nobel as "the intellectual Olympics." His successor, Carl-Henrik Heldin, made a similar analogy in the opening speech of the 2016 ceremony. The metaphor of sports competition also emerged in the headline of an English newspaper about the Nobel Prize for chemistry to Harold Kroto: "Kroto scored one for Britain."

Only one person has succeeded in entering the history of both the Nobels and the Olympics: Philip Noel-Baker, the British silver medalist for 1,500 meters at the London Olympics in 1920, awarded the Nobel Peace Prize in 1959. Coubertin was a candidate for the Nobel Peace Prize as early as 1936. But his approval of the Berlin Olympic Games in Nazi Germany, described by his supporters as an example of the conciliatory role of sport in international relations, hurt his candidacy. That year, the prize went to the pacifist and German prisoner Carl von Ossietzky in a decision that, as we shall see, led the Nazi regime to boycott the Nobel Prizes.[18]

National pride has also been evoked by the successes and international recognition linked to the Olympics and the Nobel Prizes. In 2008, when Rudi Balling announced that Berlin would be hosting the international congress for genetics in the presence of six Nobel Prize laureates, he declared: "We are as happy as if we had brought the Olympics to our country."[19]

The association of the Nobel Prize and the Olympic Games is particularly strong, as we already mentioned, in countries such as China:

> A Nobel prize, an Olympic medal or the role of Olympic host, an Academy award, a prestigious position at a famous university—recognitions like these are part of the regular news cycle in North America and Europe. But for countries on the long path upward, they can be significant, often too significant, milestones.
>
> The quest to produce a home-grown Nobel Prize for science has become entwined in China's resurgent nationalism. Having hosted a glittering Olympic Games and taken its astronauts into space, China sees a triumph on the global scientific stage as a way to convince the world that it has moved from the periphery to the centre.
>
> China is applying its strategy for winning an Olympic gold to its science policy. [...] Thus, the state is now seeking to replicate what it has learned in another domain of catch-up: Olympic sports. [...] Given the success of what is often referred to as the Chinese Communist Party's "Olympic strategy," why not develop a comparable "Nobel Prize strategy"?[20]

THE NOBEL PRIZE IN TIMES OF WAR, ACT I: TO AWARD OR NOT TO AWARD, THAT IS THE QUESTION

On October 4, 1914, the so-called Manifesto of the 93 was made public. Ninety-three German intellectuals signed it, among whom were some of the greatest scientists of the time. In fact, most of the German Nobel laureates to date were signers: the Nobel laureates for physics Wilhelm Röntgen (1901), Philipp von Lenard (1905), and Wilhelm Wien (1911); the Nobel laureates for chemistry Adolf von Baeyer (1905) and Wilhelm Ostwald (1909); and the laureate for medicine Paul Ehrlich (1908). The manifesto responded with great resolve to criticism of Germany and defended the invasion of Belgium and German militarism,

> without which German culture and civilization would have been wiped out long ago [...] we shall go ahead in this war as a civilized nation for which the

ties with Goethe, Beethoven, and Kant are as sacred as those that we feel for our homes and our families.[21]

Ostwald even made the journey to Stockholm to ask his colleagues to help spread the "truth" about Germany. The general successes of German scientists and their winning of Nobel Prizes, he claimed, were products of "German's race genius for organization."[22]

The rebuttals arrived immediately. Sir William Ramsay, British Nobel laureate for chemistry in 1904, wrote in the journal *Nature* that

> German ideals are infinitely far removed from the conception of the true man of science; and the methods by which they propose to secure what they regard as the good of humanity are, to all right-thinking men, repugnant.[23]

The Swedish Academy, the Royal Swedish Academy of Sciences, and the Karolinska Institute called an urgent joint meeting for October 20. They decided to ask the government to postpone all decisions on the prizes and to delay the prize-giving ceremony to June 1916 for all prizes for 1914 and 1915. The government authorized the delay, but the physics and chemistry departments had already chosen their honorees: a German, Max von Laue, and an American, Theodore W. Richards, respectively. Academics found themselves as divided as the public. Some felt it would be sensible to delay the prizes, perhaps still hoping that the military conflict would end quickly; others insisted that the awards proceed as usual and that "neutrality" and "passivity" should not be confused with one another. The latter position was often associated with support and sympathy, whether more or less explicit, for Germany, whose scientists were considered to be among the strongest candidates for the prizes.

In 1915, the Nobel Prize machine forged ahead, albeit at a slower pace, with its deliberations. One of the most promising candidates was the British physicist Henry Moseley. Perhaps the most brilliant student of "Nobel's maestro" Ernest Rutherford, Moseley was only twenty-five when he discovered that the atomic number, not the atomic weight, was the key to understanding the periodic table of elements. But when discussions began in Stockholm about his possible prize, Moseley was far from his laboratory: he had enlisted in the army, against the advice of the conscription committee. On August 10, news arrived that Moseley had fallen during the battle of Gallipoli. For those who supported continuing the awards even in wartime, the tragedy was a reason to continue with the awards before the war made other victims among the potential deserving candidates. Others, including the chemist Svante Arrhenius, the first Swedish Nobel laureate and one of the most influential members of the Nobel committee, saw it as a reason to be even more cautious in an ever-changing world. Additionally, as the Ostwald case illustrated, there was a risk that laureates' political views might cause embarrassment when they were suddenly in the spotlight.

Eventually, the Karolinska Nobel Assembly decided to assign the prize set aside in 1914 to the Austrian medical doctor Robert Bárány, a prisoner in Russia at the time. The Royal Swedish Academy of Sciences voted by secret ballot to award the 1914 prize for physics to Laue "for his discovery of the diffraction of X-rays by crystals" and the prize for chemistry to Richards "in recognition of his accurate determinations of the atomic weight of a large number of chemical elements." The next day, they voted for the 1915 prizes. Richard Martin Willstätter, a German and signer of the Manifesto of the 93, received the chemistry prize "for his researches on plant pigments, especially chlorophyll." The physics prize went to the British scientists William Bragg and Lawrence Bragg, father and son, "for their services in the analysis of crystal structure by means of X-rays."

When the news reached him, Lawrence was at the front developing new methods to locate enemy artillery. But even under the bombs, the Nobel celebrity effect was obvious:

Today the Curé, who had seen my photo in the paper, came in and offered me a bottle of wine with his best bow, as a little present to felicitate the occasion. Generals humbly ask my opinion about things; it is great fun. [. . .] I do hope they postpone the ceremonies. [. . .] It's freezing hard now, and very misty.[24]

The ceremony was indeed delayed until an indefinite future date. Lawrence finally made his speech in Stockholm in 1922. William Bragg, shaken by the death of his second son, Robert, who was wounded in Gallipoli, declined all future invitations to celebrate his prize, not wanting to find himself at a ceremony next to scientists of enemy nations ("I imagine there will be many Germans," he confided to Rutherford).[25]

On balance, the Academy succeeded in managing the situation in a "politically satisfying manner: a neutral American prize, one to Great

Britain and two to Germany."[26] But new and even more dramatic controversies were just around the corner.

"TO THE BENEFIT OF MANKIND": NOBEL PRIZE OR WAR CRIMINAL?

During the years of the First World War, the Nobel Prizes existed in limbo. There were no ceremonies or prizes awarded in 1916 and 1917. When the war's outcome became clearer, ambitions that went well beyond the prizes began taking shape among key figures on the Nobel committees. For Arrhenius and other academics, Germany remained a model to emulate in research organization and funding, but its loss of strength and influence seemed destined to leave a void—a void that Sweden and its scientists were ready to fill, presenting themselves, thanks to the Nobel Prizes, as natural and neutral mediators between the winners and the defeated in the context of increasingly contentious and rivalrous international scientific relations. To continue advancing scientific discovery, it was essential to collaborate with scientists from allied countries, while reaching out to German scientists. Arrhenius and others dreamed of scientists from enemy countries sitting side by side and shaking hands during the Nobel Prize ceremonies and banquets.

In 1918, it was decided the Nobel Prize should keep a low profile. The 1918 prize was set aside, and the 1917 prize was awarded to the British scientist Charles Glover Barkla "for his discovery of the characteristic Röntgen radiation of the elements." He received only one nomination, but it was a very high-profile one from former Nobel Prize recipient Rutherford. His presentation was among the weakest in the history of the Nobel, but his simple experimental approach—not to mention his hostility "towards the new world of relativity and quantum theories"—were the type of physics that pleased Stockholm scientists immensely in those years. Einstein, by contrast, was thought to be too radical, although he received numerous nominations (see chapter 4). Johannes Stark's candidacy was compromised by the support he received from Lenard, one of the most active scientists in the nationalistic defense of the German cause. Some disparaged the 1918 prize for physics as "a prize to please the British."

But the real bombshell—literally—concerned the prize in chemistry. The bomb had a name and surname: Fritz Haber. In 1909, Haber had found a way to make ammonia with hydrogen and nitrogen. With the help of the expert Carl Bosch, the chemical company BASF scaled up the Haber method to an industrial level. The resulting process is known as the Haber-Bosch process and is extremely important for the production of agricultural fertilizers, which dramatically increased food production. In 1914, the BASF plant produced 7,200 metric tons of ammonia per day, which could produce 36,000 metric tons of fertilizer.

When ammonia began flowing from BASF, Europe was already at war. Rather than fertilizing crops, BASF ammonia was used by the German war industry to produce explosives. Some even claimed that without the Haber-Bosch process, the war might have ended sooner because German nitrate stocks would have run out very quickly. Furthermore, the Haber-Bosch process contributed to an imbalance between the military capacities of Germany and its enemies, as the patent omitted important technical details and prevented other nations from replicating the process.

On top of this, Haber played an active role in the conflict. In 1915, as a captain in the army, he personally coordinated the first uses of lethal chlorine gas on French soldiers, leaving five thousand victims on the ground and inflicting horrendous injuries. Considering these as experiments, he derived the formula that carries his name and established the relationship between gas concentration and mortality. Also in 1915, after years of supporting and contributing to her husband's researches, Haber's wife, the brilliant chemist Clara Immerwahr, shot herself in the chest. The day after her suicide, Haber left for another front, where a gas attack was planned against Russian enemies. We hardly need to note that he was among the signatories of the Manifesto of the 93.

Haber was nominated for the Nobel Prize in chemistry in 1912, 1913, 1915, and 1916. In 1918, his supporters on the Nobel committee insisted on putting his name forward again although he had received only one nomination. Their argument, incredibly, was that the war had demonstrated the efficiency of Haber's result, which would be beneficial for "humanity" in the coming years. The entire Academy met on November 12, 1918, the day after the armistice. After a heated discussion, it was decided that the chemistry prize for that year should not be given and should be set aside for the following year. The reasons for this were practical (there was not enough information to establish whether to reward Bosch as well) and political. A year later, the pro-Haber pressure on the Academy ramped up. Some felt it could be Haber's last chance: rumor had it that his name had been added to the list of war criminals put together by the Allies and that he was hiding in Switzerland, attempting to disguise himself by growing a beard. In the end, the Academy awarded the Nobel Prize for chemistry to Fritz Haber "for the synthesis of ammonia from its elements." As for physics, the 1918 and 1919 prizes were a German "double whammy"; they were awarded to Max Planck (also a signatory of the Manifesto, although he later changed his views) and Johannes Stark, who had been passed over the previous year due to concerns about his closeness to nationalistic German groups.

Controversy over Haber's recognition in particular exploded in the press and in public opinion. French papers called it "the Nobel Prize scandal" and "the glorification of barbarity." For the Swedish paper *Dagens Nyheter*, the prize to Haber was

a formal declaration of war by the neutral and above the turmoil of battle elevated Swedish science.[27]

The Social Democrat Hjalmar Branting, who had initially objected to the institution of the prize itself (but was eventually awarded the Peace Prize himself), asked the controversial question of what might have happened if a French or British chemist working with the military had been rewarded

while the sores are still dripping blood? But because Haber is a Prussian professor we find it acceptable![28]

The Swedish conservative press defended Haber's prize, touting the "objective merits of German science." The members of the Academy claimed to pass impartial judgment and that political considerations were not their domain. In his reply to the letter announcing the prize, Haber did not miss the opportunity to underline that the prize was

a tribute to the accomplishments of German science from professional colleagues in Sweden who, as much by their scientific eminence as by their neutral position between the larger countries, have been called upon to exercise a superior and impartial judgment.[29]

The controversy extended also to the Italian press:

The Nobel Prize was awarded to the German Haber, known especially for his work on the industrial use of atmospheric nitrogen. If his only claim to fame were the progress from which agriculture and industry will greatly benefit, there would be nothing to criticize. However, his winning the prize has given rise to a very lively debate across the scientific world and in Sweden in particular. The reason being that Haber is the creator of asphyxiating gas: of the terrible German asphyxiating gas. Was the Swedish Academy unaware of this? And if it was aware, did it not understand that by rewarding Haber, it was denying its prizes of their high value for the rewarding of the noblest efforts toward progress?[30]

This year's prizes have been monopolized by German science: among the winners who came to Stockholm were five German scientists, one of whom was professor Haber, the inventor of asphyxiating gas. The father of this deadly invention was thus awarded a sum of more than one million German marks: a paradox that requires no further comment.[31]

After years of being postponed, the Nobel celebrations were held in the beginning of June 1920. However, this was not a day of hugs and handshakes as dreamed by Arrhenius and the other supporters of a "politico-scientific" reconciliation between Germany and the Allies. On a cold rainy day in Stockholm, the only non-German in attendance was Barkla: the Braggs and Richards chose not to attend. And the royal family was not present either, because it was in mourning after the passing of Princess Margaret. The president of the Nobel Foundation handed out the prizes.

When the prizes were awarded, Haber's name appeared both on the Nobel Prize honor roll and the list of war criminals (although he was never formally tried). He would go on to maintain a good relationship with the Academy of Sciences and sent numerous nominations for some of his colleagues. Five other scientists who had worked on the gas, some of whom collaborated directly with Haber, went on to win Nobel Prizes in following years: Walther Nernst, Heinrich Wieland, James Franck, Otto Hahn, and Gustav Hertz.

Even his contribution to the German military effort in World War I and the glory conferred by the prize did not protect Haber from being expelled from Germany under Nazism due to his Jewish origins. When Max Planck asked Hitler for clemency on Haber's behalf, the Führer answered: "If science can't manage without the Jews, we will have to manage without science for a few years." Between the two wars, having left Germany, Haber continued his research. An insecticide that he created was used to produce Zyklon B, a gas used in some concentration camps to murder prisoners.

THE NOBEL PRIZE IN TIMES OF WAR, ACT II: NO POLITICS, PLEASE, WE ARE SWEDISH (SCIENTISTS)!

In the mid-1930s, clouds again gathered above Stockholm. This time, the storm came from Norway. In 1936, the Norwegian Parliament's Nobel committee for peace disregarded, among others, the candidacy of Pierre de Coubertin, the inventor of the modern Olympics, and decided to reward the German journalist Carl von Ossietzky (with a prize held over from the previous year). Ossiestzky was a fierce opponent of the Nazi regime and a concentration camp prisoner. The Nazi regime responded by immediately boycotting anything linked to the Nobel Prize.

The boycott did not discourage several members of the Academy of Sciences from maintaining strong and respectful working relationships with German science. Hans von Euler-Chelpin, a German scientist and a professor in Stockholm, proved a key figure. He had received the Nobel Prize for chemistry in 1929 for "his investigations on the fermentation of sugar and fermentative enzymes." He hoped to become a key player in German research, even if it meant compromising with Nazis. Once he acquired some influence within the Academy, a Nobel Prize appeared to be an ideal instrument. Euler-Chelpin therefore campaigned on behalf of his German colleagues, convinced, on one hand, that the Nazi regime could not afford to miss out on a Nobel Prize as a propaganda instrument and, on the other, that the prize would risk losing credibility without German scientists.

In 1938, the Academy again decided not to award any prizes. The following year, two of Euler-Chelpin's favorites were rewarded: Richard Kuhn (who was awarded the prize for chemistry that had not been assigned in 1938) and Adolf Friedrich Johann Butenandt (who received the prize for 1939 and divided it with the Swiss Leopold Ruzicka). In addition to Gerhard Domagk, who won the prize for medicine, the total of German laureates came to three. For the Nazi regime, it was a provocation; for most German scientists, it was a decision that risked being more harmful than beneficial. All three German laureates officially refused the prize. Kuhn and Butenandt later said they had been summoned by the Education Ministry and handed two letters declining the awards. Domagk, whose discovery had truly extraordinary implications (he proved the efficiency of sulphonamides in antibacterial therapy after experimenting on his daughter when she was affected by a serious infection), was arrested by the Gestapo and released only after he signed the letter of renunciation. In 1947, he went to Stockholm to receive his diploma and medal; the other two received their medals and diplomas from the Swedish consul. None of them ever obtained the financial reward.

The Nobel Prizes were again put on hold during World War II, resuming in 1944. The tragedies of Hiroshima and Nagasaki at the end of the war made even more clear the decisive role that scientists could play.

At this point, one could expect the Haber case and all its controversies to have served as a lesson to the Nobel committees. But a similar mix of science, politics, and personal ambition came into play in the 1940s. The German scientist Otto Hahn, one of the fathers of nuclear fission, and his colleague Lise Meitner came to the attention of the Nobel committee for chemistry. Recognizing fission meant rewarding an important scientific (and military) achievement but also signaled support for the ambitions of certain members of the Academy—such as Karl "Manne" Siegbahn, who aspired to lead the Swedish atomic program. As in other times, at least according to some, it reflected the independence of the Swedish Academy: rewarding a German was proof of not being dominated by the Americans. Sweden was also looking ahead to postwar collaborations with Germany by elevating a scientist not linked to Nazism.

The 1944 prize was set aside to avoid a Nazi boycott, and it was given to Hahn in 1945. When the first rumors about the prize started circulating, the British magazine *New Statesman* ironically commented that Hahn should be awarded the Nobel Prize both for chemistry and for peace, considering "he knew the secret to make the atomic bomb and didn't pass it on to Hitler."[32]

In the meantime, Hahn had disappeared. It was soon discovered that he was being held prisoner at Farm Hall in England with nine other scientists suspected of having taken part in the German atomic program—an unfounded suspicion in his case. Hahn and his fellow prisoners received the news of the Nobel Prize with songs and celebrations. Another imprisoned German scientist (and prior Nobel laureate), Max von Laue, said to Hahn,

You have got the Nobel Prize as a consolation for we ten German scientists who are shut up here, but I think that is their reason for giving it to you now instead of waiting till next year, which was probably their original intention.[33]

Hahn would have to wait until December 4. He was permitted to inform the Academy that he accepted the prize; however, he was not allowed to travel to Stockholm to receive it until the following year. No prizes were given to Hahn's colleague, Lise Meitner. We will look further into her dramatic story in chapter 5.

The 1945 prize in chemistry, to Finnish scientist Artturi Ilmari Virtanen, was also perhaps politically motivated. The award "for his research and inventions in agricultural and nutrition chemistry, especially for his fodder preservation method" came as a rather surprising choice, supported by very few nominations. However, it honored scientists in a country that had been badly affected by the war against the Soviet Union and that many Swedish academics (Euler-Chelpin among them) saw as an important ally and rampart against the Bolsheviks.

THE (STOCKHOLM) SYNDROME OF THE PLATYPUS: IS THIS CHEMISTRY, PHYSICS, OR MEDICINE?

"Physics is the only true science: the rest is just a stamp collection." The great physicist Ernest Rutherford liked to settle arguments between disciplines with this scornful remark. These were the first years of the twentieth century, in which physics and chemistry competed, at times violently, for phenomena and objects of study.

The boundaries set by the Nobel Prizes, in fact, were not only national but even disciplinary. Chemistry, physics, and medicine—the three disciplines recognized by the prizes—saw their boundaries shift over time according to the situation.

A little like the troubled history of the classification of the platypus, which for nearly a century, from 1798 to 1884, continued to slide among various classes of animals. The beak, the mammary glands, the presence in aquatic environments, the hair, and the egg laying were differently emphasized by naturalists who supported one or the other classification.[34]

Let's do a little test. Marie Curie won a Nobel Prize in 1903 with her husband Pierre Curie "in recognition of the extraordinary services they have rendered by their joint researches on radiation phenomena discovered by Professor Henri Becquerel" (awarded for "his discovery of spontaneous radioactivity"). Chemistry or physics? Physics. In 1911, Marie Curie, one of the very few to have won two Nobel Prizes, again won a Nobel Prize "for her discovery of the elements radium and polonium." Chemistry or physics? Chemistry, this time. Same scientist, same field of research, two different prizes in two different subjects. Let's look at another case that we saw earlier: Otto Hahn's 1944 Nobel Prize (awarded in 1945) for "his discovery of the fission of heavy nuclei." This seems like a physics prize, but Hahn was awarded the prize for chemistry. Many recipients in physics were also nominated for chemistry and vice versa. Emilio Segrè, a physicist of Italian origin who was awarded a prize in 1959 "for the discovery of the antiproton," was nominated nine times for chemistry; and Enrico Fermi, the Nobel Prize for physics in 1938, was nominated three times for chemistry too.

At the origin of these fluid disciplinary boundaries there are some general reasons and others more contingent. Particularly during the first years of the prize, the status of so-called chemical physics was uncertain, and competition over "objects of research" such as the atom was fierce. The new field and new research made it "difficult to determine where chemistry ended and physics started."[35]

The contingent reasons had to do, as in other cases, with personal dynamics within the Academy. The key figure, in this case, is again that of Svante Arrhenius. A massive, ruddy figure, with a combative and volcanic personality, a swirling moustache, and a bow tie, Arrhenius had many scientific interests and intuitions. He was one of the first to calculate the relation between carbon dioxide and the increase in temperature of the earth (later known as the "greenhouse effect")—in 1896!

His doctoral dissertation was almost rejected by the chemistry professors of the University of Uppsala, but his work was much admired by influential foreign colleagues in the emerging field of chemical physics.

His name circulated immediately as a possible candidate when the Nobel Prize was created. In 1903, the Nobel committee for chemistry made a strange proposal to the committee for physics. Since the Academy voted jointly on the final decisions of both prizes and since Arrhenius had received nominations for both subjects, why not give him half a prize for physics and half for chemistry? The Nobel committee for physics rejected the idea and granted their award to the Curies and Becquerel. The chemists of the Academy were not pleased, as they

> saw the award of the physics prize for pioneering work in radioactivity as a dangerous precedent, since it could mean that in the future this field would be appropriated by the physicists for their prize.[36]

Giving a prize in physics for radioactivity was a way of staking a claim over a rising field of study. Some chemists already felt they were being colonized by physics and that their own subject's "glorious future was now behind them,"[37] after their numerous successes in the previous century. Between 1901 and 1930, works in chemical physics received nine prizes, a significant proportion of the total.

Arrhenius received the Nobel Prize for chemistry in 1903 after a very tight vote. His position as the first Swedish Nobel Prize recipient granted him influence in the committees' future choices and navigations among the fluid boundaries between subjects during those years. Rewarded for chemistry, he became a member of the Nobel committee for physics but continued to send nominations for chemistry, too, without ever forgetting that it was the Swedish chemists who had risked halting his brilliant career before it had even taken off.

Throughout the Cold War, Swedish academics maintained lines of communication with scientists from the Soviet Union. Although Soviet winners of Nobel Prizes for peace and literature were prevented from accepting their awards, all chosen Soviet scientists were able to go to Stockholm to receive their prizes. In November 1955, however, the leadership of the Section of Physical, Technical, and Mathematical Sciences of the Soviet Academy of Sciences adopted a resolution by which

> it does not find it advisable to nominate Soviet scientists for the Nobel Prize since this prize cannot be considered international as demonstrated by the lack of Nobel awards to outstanding individuals of science and culture of our country (D. I. Mendeleev, L. N. Toltstoi, A. P. Chekhov, M. Gorkii).[38]

Though this may seem directed to an international audience, it should instead be read as a dispute between Soviet physicists and mathematicians on one hand and chemists on the other. The physicists were aware they had no potential winners during those years, while there was a rumor of a prize in chemistry for Nikolaj Semenov (who was indeed the first Soviet scientist to be rewarded, in 1956).

The prize for physiology or medicine—awarded by a different institution, the Karolinska Institute—was largely absent from these "boundary" skirmishes. However, with the advent of biochemistry, some issues arose between chemistry and medicine. The prize to Francis Crick, James Watson, and Maurice Wilkins for the discovery of the structure of DNA in 1953 is the obvious example.

Watson and Crick were nominated seven times for chemistry, once in 1960 and six times in 1962, and ten times for medicine (two in 1960, three in 1961, and five in 1962). Lawrence Bragg, the former *enfant prodige* who had received the physics prize together with his father at a very young age, orchestrated their candidacy, partly to increase the stature of his Cavendish Institute.

After Rutherford's death, Bragg hoped to steer the Cavendish Institute in a new direction, focusing on the new frontiers of biology instead of physics. Crick himself had started his career in physics, but a bomb destroyed all his equipment during the war. Watson, on the other hand, had a doctorate in zoology.

To have a Nobel Prize, actually two, would be for Bragg an affirmation of that new strategy. Using all his influence and counting on the fact that he was himself a celebrated Nobel prize, he nominated Watson, Crick, and Wilkins for chemistry and two other scientists from Cambridge, Max Perutz and John Kendrew, for physics. It was a hand of four aces for the Cavendish Institute, plus recognition of the role played by colleagues (and rivals) from King's College with the nomination of Wilkins (who also had a background in physics). But it became apparent that the candidacies of Perutz and Kendrew were better suited to chemistry. In 1962, the biochemist Arne Tiselius—Nobel Prize laureate for chemistry in 1948, member of the committee for chemistry, and a very influential figure in the Nobel Prize world—proposed a possible solution: Why not award the prize for chemistry to Perutz and Kendrew and the one for physiology or medicine to Watson and Crick?

The Karolinska committee leapt at the opportunity to be associated with such a notable discovery and to make amends for their omission of Oswald Avery's earlier discovery (in 1945) that DNA, not protein molecules, transmitted genetic information. Despite Avery's thirty-eight nominations, the committee had not awarded him the prize before his death in 1955.

The Karolinska waited for the Academy of Sciences to announce the Nobel Prize for chemistry to Perutz and Kendrew on October 11, 1962. It then voted unanimously to award the Nobel Prize for medicine to Watson, Crick, and Wilkins on October 17.[39] In this way, two physicists and a zoologist, who had been nominated also for chemistry, received the Nobel Prize for medicine.

There is another reason to be interested in the Nobel Prize. The history of the prize is an opportunity to study the forces at play in the collaboration and competition between scientific subjects and their evolution over the years. In doing so, it is useful to focus on individuals as well as the relations between fields of study. The prize categories as stipulated by Alfred Nobel's last wishes have been interpreted and defined differently over history (like the definition of the platypus by naturalists) according to the development of the different subjects, their cultural relevance, and the priorities—without going so far as to say the "preferences"—of the Nobel Prize juries.

In the medical field, for example, in the first decades of the prize, physiology and bacteriology won most of the prizes. Discoveries in bacteriology alone won nearly half the prizes. But between 1933 and 1995, the comparatively new fields of biochemistry and genetics commanded an impressive twenty-six prizes in medicine, with advances in genetics garnering an additional ten prizes in chemistry. For example: the prize in 1980 to Paul Berg ("for his fundamental studies of the biochemistry of nucleic acids, with particular regard to recombinant-DNA"); to Walter Gilbert and Frederick Sanger ("for their contributions concerning the determination of base sequences in nucleic acids"); the prize in 1989 to Sidney Altman and Thomas Cech ("for their discovery of the catalytic properties of RNA"); and the 1993 prize to Kary Mullis and Michael Smith ("for contributions to the developments of methods within DNA-based chemistry").

Meanwhile, developments in the field of surgery were largely neglected, despite the "benefit to mankind" noted by the prize's founder. Only two prizes have been awarded for the development of a new surgical technique. The first, in 1909, went to Emil Theodor Kocher. The 1912 prize in medicine went to Alexis Carrel, a physician of French origin, "in recognition of his work on vascular suture and the transplantation of blood vessels and organs." Surgical applications were occasionally mentioned in other prizes but rarely honored as the primary triumph. In substance, the Nobel committee has seemed to interpret "medicine" as biomedical research.

In chemistry, we have mentioned the supremacy of chemical physics and biochemistry. In physics, important fields such as astrophysics and geophysics were excluded for many years. Arrhenius (him again!) supported the idea that astrophysics belonged to the field of astronomy and even tried, unsuccessfully, to change the Statutes of the Nobel Foundation to officially exclude astrophysics from any possible prizes. It took almost forty years to achieve recognition in this field (in 1935, Victor Hess won the prize "for his discovery of cosmic radiation") and another thirty for the decision to reward the theory by Hans Bethe on active nuclear processes in the sun (he was awarded the prize in 1967 for an explanation dating back to 1938). Edwin Hubble, Arthur Eddington, and Alfred Wegener are key twentieth-century scientists whose names are absent from the prize's roll of honor. The first prize for geophysics was awarded only in 1947 to Edward Appleton for his studies on the ionosphere, which dated back more than twenty years.

Generally speaking, especially during the first decades, the Nobel committee considered physics to mean experimental physics. Despite numerous nominations for results in theoretical physics between 1901 and 1920, only four theoretical physicists were recognized by the prize committee. Inventions, although specifically mentioned in Nobel's will, have received even less laudatory attention: overall, few were awarded, among them that by Marconi in 1909.

This changing and often contingent definition of the different subjects and their boundaries has been humorously noted over time. In research circles, a quote by Arthur Bloch has earned frequent mention and a spot on the walls of some laboratories: "If it is green or it moves, it's biology.

If it stinks, it's chemistry. If it doesn't work, it's physics." When Rutherford's colleague Frederick Soddy commented about their experimental results, "Rutherford, but this is transmutation!," Rutherford famously replied: "For Mike's sake, Soddy, don't call it transmutation. They'll have our heads off as alchemists."[40] Since we began with Rutherford's disdain for everything but physics (all of which was, according to him, "stamp collecting"), we shall finish with him as well. Ernest Rutherford won the Nobel Prize in 1908. For which discipline? Chemistry, of course.

4

AND THE WINNER IS . . . NOT HERE! HOW EINSTEIN WON THE NOBEL PRIZE AND WHY HE ALMOST NEVER RECEIVED IT

This invisibility, in fact, is only good in two cases: It's useful in getting away, it's useful in approaching.
—H. G. Wells, *The Invisible Man*

Close your eyes and imagine the Nobel Prize. Which scientist comes to mind, and which invention do you see? You will most likely have thought of Albert Einstein and his theory of relativity. Right? And you probably envisioned him with his famous bush of unkempt hair, picking up his prize from the hands of the king of Sweden with a bow and a smile.

Seems reasonable, but it's a shame that this scene never actually took place.

Of all the fascinating stories associated with the prize, it's difficult to land on one that better illustrates the intrigues of science, society, and politics better than the one of the Nobel Prize to this great physicist. The time has come to tell the story.[1]

STOCKHOLM, 1922: A WALK ALONG THE WATERFRONT

At the end of May on the Strandvägen seafront, spring smells like a summer promise.

Carl Wilhelm Oseen feels no promise, however, nor does he smell springtime. The season is passing by without his notice, like a small glittering fish slipping along the shoreline. He walks back up the Strandvägen, head bowed, meandering up the paths along the shore, possibly trying to avoid thinking about his problems or rather *the* problem that had been tormenting him for some time. "Maybe I shouldn't have accepted this job," he mumbles to himself repeatedly, lips barely moving.

As he walks, he can hear the waves shifting the pebbles on the shore and the swishing of the small boats as they cut through the water. Carl Wilhelm's specialty is the slow movement of small particles in fluid—small movements that make all the difference.

As a child, Carl Wilhelm liked to drop tiny stones or particles of sand into a glass and observe their dancing through the water, trying to guess how they would move before reaching the bottom. Newton liked to describe himself as a small child playing intently on the beach, happy to find a pebble that was flatter or whiter than the others; Carl Wilhelm was fascinated by their minuscule movements.

Was there any pattern, or did each tiny stone move through the fluid in a completely random way? If we are able, today, to dose the precise diffusion of some drugs through the blood or even understand how smog particles behave, it is in part thanks to him.

On this day, however, he can feel the viscosity and inertia that he knows so well seeping through him, as if he himself were a particle dragged down by a dense dark liquid with no possibility of changing his path.

And then, suddenly, he sees light. A ray of sunshine breaks through the trees, and Carl Wilhelm can't help looking up as the ray reflects onto the metal sign of the coffee shop on the pier. Light, therefore. But of course. For Carl Wilhelm, this solves everything. The particle starts moving through the liquid once again. Words like *justice* or *truth* are too abstract for a practical and pragmatic mind such as his. But even a tiny particle can contribute, in certain circumstances, to things going in a certain direction.

Carl Wilhelm turns on his heels, leaving the sea to his left. He increases his pace until he reaches his office on Sveavägen. He practically runs up the wooden steps. It is Sunday, and there is no one in the office. He searches through a few large books on the shelves in the library and takes out one dated 1905. He opens it, skims the index, and turns straight to the page he is interested in. Then he closes the book. Perfect. Yes, that might work.

To stay there and work is not safe. Best to go home. Now that he knows what to do, it won't take long. He would like to take the book home, but to leave an empty gap at that precise place in the library would raise suspicions. Standing up, without even taking off his jacket, he jots down a few things in his notebook. After months of tension, he finally smiles.

BERN, 1905: AN EMPLOYEE'S "SECRET" OFFICE OF THEORETICAL PHYSICS

From the outside, it's hard to believe that history was made in this austere building on the corner of Speichergasse and Genfergasse.

Albert works here as a "third-class technician." He got the job thanks to a recommendation from Marcel Grossmann, having been turned down for assistant positions by various German, Swiss, Dutch, and Italian universities. It must be said that his doctoral thesis had been a small ordeal: he handed it in late, and a few mistakes were subsequently corrected.

Albert's work consists mainly in evaluating patent requests. A substantial part of these requests has to do with the measuring of time: watches and particularly systems for the regulation and synchronization of watches. Coordinating time in Europe, where rail transport is rapidly developing, is a hugely important problem. Increasingly sophisticated chronometers—devices to allow time signals to travel through telephone wires or even wireless systems—keep piling up on Albert's desk.

His grumpy supervisor, Friedrich Haller, has warned him to be careful and avoid agreeing with every inventor's way of thinking. Always start from the assumption, he tells Albert, that everything written in the patent request is wrong.

Albert tries. He spends eight hours a day in the office, six days per week. He often goes to the Café Bollwerk near the station for lunch, passing by the large clock at the entrance that monitors the punctuality of departing and arriving trains. It is one of the numerous coordinated electric clocks that the city of Bern is so proud of. Another, possibly more beautiful one sits on the Kramgasse Tower. Albert sees it every morning as he turns left toward the patent office after leaving home. He sees it in the evenings, too, when he returns home, slightly depressed at the thought that another day has gone by without having been able to develop a certain idea that he has been toying with for a while. Furthermore, when he leaves the office, the university library would already have been closed for a while, and he would not be able therefore to read any recent works by colleagues who—who knows—might have helped him clarify a few points.

He lives with his wife, Mileva, and their two children in Kramgasse in a modest two-room flat on the second floor of number 49. While Mileva cooks their dinner, Albert enjoys constructing small trains out of matchboxes, much to his youngest child's delight—the famous trains for which inventors across Europe are striving to coordinate the movements with the most sophisticated chronometers. He continues to study every evening at the living room table under the large grandfather clock. And he probably studies in the office, too, between one patent request and another, just like your average employee who cultivates a passion that he hides from his boss. When a colleague visits him, he opens his desk drawer and tells them jokingly that this is his theoretical physics office.

When he applies for a promotion to second-class technician, his boss turns him down. Albert has made progress, he reports, but still has to acquire skills in mechanical engineering.

One day, during that same spring, he is taking a walk with his best friend, Michele. They talk about music, gossip about the personnel in the patent office, but especially speak about electromagnetism. And then about the problem—the real problem—that Albert has been working on since the age of sixteen.

Michele has to repeat his last question because Albert is not answering him. He turns around but can no longer see his friend. Albert has stopped under the portico and is standing there, still and silent, staring into thin air. He slept very badly the previous night. The next day, when he meets Michele, he thanks him. He has completely solved the problem, Albert says. Then he leads Michele to a hill to the northeast of Bern. From there, he points to Bern's clock tower and then to the one in Muri, a little village loved by Bernese aristocrats.

Albert goes home and writes to another friend, Conrad. "I will soon send you four new articles, one of which will be of great interest to you," he says. The articles he refers to consist of thirty pages of text and formulas that would be published in the *Annalen der Physik* journal in June 1905 under the title "On the Electrodynamics of Moving Bodies," the starting points of which—needless to say—are trains and clocks:

> If, for example, I say that "the train arrives here at 7 o'clock," that means, more or less, "the pointing of the small hand of my clock to 7 and the arrival of the train are simultaneous events.[2]

A few weeks pass, and Albert continues to think about his article. He has a feeling something is missing, something that is the inevitable conclusion of those thirty pages. Thus he writes a three-page postscript. In those three pages, he lays out the theory that would later be summarized in the equation that we have all seen at least once in our lives, $E=mc^2$—although the letters he uses originally are different:

> If a body releases the energy L in the form of radiation, its mass decreases by L/V^2.[3]

And he concludes cautiously:

> Perhaps it will prove possible to test this theory using bodies whose energy content is variable to a high degree (e.g., salts of radium).[4]

During that *annus mirabilis*—or rather during that *mirabilis* spring/summer—Albert writes two other articles. One is his doctoral thesis. It examines a new way of determining the sizes of molecules. It is considered the less important of the two, but for a long time has been the most frequently cited by his scientist colleagues. The fourth article is titled "A Heuristic Point of View about the Creation and Conversion of Light," and it was published on pages 132 to 148 of the same volume of the *Annalen der Physik*. Light, precisely.

EVERY GREAT ENTERPRISE NEEDS AN ENEMY

Every great enterprise needs an enemy, and Carl Wilhelm Oseen's enemy was Allvar Gullstrand. If one looks at Gullstrand's photograph today, his hard gaze is slightly daunting. But that could be in hindsight.

Allvar also deals with light, in a way. His work is fundamental in understanding the refraction of light in the eye. If you go for an eye test, to this day the instrument the doctor uses to look into your eyes is based on one of his inventions, the slit lamp. In 1911, Gullstrand receives the Nobel

Prize for medicine or physiology. He also was put forward for the prize in physics. He prefers the prize for medicine but then gets himself elected to the Nobel committee for physics. Proposals from colleagues around the world are sent to the committee, which evaluates and assigns the proposals that are then voted on by the Academy of Sciences.

From 1901 onward, the committee receives numerous proposals for the Nobel in physics. One name is (almost) always among them: that of Albert Einstein. The name is put forward for the first time by the chemist and physicist Wilhelm Ostwald, the same man Einstein contacted unsuccessfully years earlier for an assistant position. The committee is not convinced by the theory of relativity and chooses to wait for further experimental developments and confirmations.

One day in November 1919, however, things change, or at least they seemingly change. Einstein wakes up to find his name on the front page of the major international newspapers, among them the London *Times* and the *New York Times*. The papers report on the results of new astronomical observations that confirm the theory of general relativity during a total eclipse of the sun. "Lights All Askew in the Heavens, . . ." reads the *New York Times*: "Einstein Theory Triumphs." The physicist has become a world-famous celebrity, and the media follow him everywhere for a comment or a photo.

Well, almost everywhere. In Germany, not everyone is celebrating Einstein's success. In 1920, a group of physicists, among whom are the two Nobel Prize recipients Philipp von Lenard and Johannes Stark, mobilize explicitly against Einstein and relativity, which they consider to be an emblem of a purely speculative form of physics, far from experimental practicality and dangerously in line with the revolutionary political and artistic upheavals of those times. This anti-relativity and anti-Einstein movement is linked at times to feelings of nationalism and anti-Semitism. Beginning in 1920, one agitator and anti-Semitic activist, Paul Weyland, organizes a series of lectures at the Berlin Philharmonic against relativity that he defines as "scientific Dadaism."

In the meantime, the proposals to award the Nobel Prize for physics to Einstein keep pouring in. In 1921, his name appears fourteen times out of the twenty-one names that are put forward, and many nominations refer explicitly to his theory of relativity. But Swedish academics remain perplexed, possibly even annoyed, by the excitement that has taken over Sweden for a theory that is so complicated to understand. It is possible that some of them find it difficult to picture Albert, with his unconventional character and disheveled hair, attending the austere and formal Nobel ceremony. But at this point, something has to be done. That is when Allvar makes his entrance. He takes it on himself to write a report on relativity. He cannot understand some of the ideas and has to ask Carl Wilhelm to explain them to him, seeing as he is more knowledgeable from a theoretical point of view. He objects continually to the theory, but Carl Wilhelm refutes each objection patiently. He finally produces a fifty-page document in which he basically dismisses relativity as mere speculation, more metaphysical than physical. When the Academy of Sciences meets, most colleagues are inclined to agree with Allvar and his judgment on Einstein's theories. Some hesitantly object that the case might become an embarrassment to the Academy; after all, the scientist is now more famous than the Nobel Prize itself. But Allvar abruptly puts an end to the discussion: "Einstein must never win a Nobel Prize, even if the whole world demands it!" The Nobel Prize for physics is not awarded to anyone in 1921 and is kept aside for the following year.

In the meantime, the situation in Germany is worsening. In spring, Albert is criticized for visiting France, the country's historical enemy. "By now, he is relativizing even our feelings and national honor," they say of him.[5]

On the morning of June 24, Foreign Minister Walther Rathenau, Jewish and a friend of Albert, is driving from his home to the ministry. A car pulls up beside him, and gunshots are fired into his car. The killers are stopped and found to be two former army officers. Albert is warned that he could be one of the next victims.

On July 6, 1922, he writes to Max Planck:

Dear Colleague: This letter is not easy for me to write, but it really does have to be done. I must inform you that I cannot deliver the talk I promised for the Scientists' Convention, despite my earlier definite commitment. For I have been warned by some thoroughly reliable persons (many of them, independently) against staying in Berlin at present and generally particularly against making any kind of public appearances in Germany. For, I am supposedly among the group of persons being targeted by nationalist assassins. I have no secure proof, of course; but the prevailing situation now makes it appear thoroughly credible.

If it had been an action of substantial professional importance, I would not have let myself be swayed by such motives, but a merely formal act is involved that someone (e.g., Laue) could easily perform in my place. The whole difficulty arises from the fact that newspapers mentioned my name too often and thereby mobilized the riffraff against me. So there is no helping it besides patience and—leaving town.[6]

In autumn, he is expected to be the guest of honor at the annual convention of an important scientific association in Lipsia. But on August 5, a threatening article is published in the city's local paper protesting against him and relativity, undersigned by numerous physicists, among them Lenard and Gehrcke. On August 8, another newspaper headline reads "Einstein auf der Mordliste": Einstein on the list of the next assassinations. At this point, Einstein has abandoned the idea of attending the conference in Lipsia and reduced his public appearances to a minimum.

STOCKHOLM, 1922: CARL WILHELM'S SECRET PLAN

Carl Wilhelm arrives at the meeting of the Nobel committee for physics on May 30.

The previous year, he gained his colleagues' trust partly thanks to his patient explanations to Allvar. The wisest among them are conscious of the importance of his expertise in theoretical physics: since they need his counsel, they know they might as well include him from the start.

The others are not aware that Carl Wilhelm has a plan—the only plan that will allow him to overrule Allvar and his categorical opposition to relativity.

The meeting goes exactly as he has foreseen. They give Allvar the task of updating the document on relativity. Carl Wilhelm is asked to present the theory of the photoelectric effect, or rather "the discovery of the *law* of the photoelectric effect." One would not want the term *theory* to sound too speculative. On July 22, 1922, Allvar hands in his report, and unsurprisingly, it is another firm dismissal of relativity—possibly harsher than the previous one. Soon after, Carl Wilhelm hands in his report, and this time, no one dares to object. It is decided that Albert should be awarded the Nobel set aside in 1921 for his "services to theoretical physics, and especially for his discovery of the law of the photoelectric effect."

As soon as the news becomes public, Sweden is bubbling with excitement for Albert's arrival. Some shudder, however, at the news that the agitator Paul Weyland is also due to visit Sweden. Is this an attempt to foment the more extremist fringes against Albert's arrival? Or a plan for a spectacular attack during the awards ceremony?

In the end, however, Albert is not present at the ceremony in Stockholm, which takes place on the anniversary of the death of its founder, Alfred Nobel. His chair, opposite the king of Sweden, stays empty. The news of the prize reaches him while he is traveling on a steamship to Japan, where he will spend a few very pleasant months, far away from the preoccupations that assail him back home. He is gratified by the prize, which also gives him some financial respite, notwithstanding the fact that he agreed at the time of his separation to make the money available to his former wife, Mileva, in the event of him ever receiving the prestigious and significant award from the Academy of Sciences. It is possible that some members of the Academy are also relieved by his absence. Who knows how the impatient character would have reacted when faced with the Nobel diploma, the only one in the whole history of the prize that contains a sort of "warning notice"?

Relativity is still problematic for some of the physicists on the committee, who therefore decide that Albert will receive the prize, as stated on the diploma, "independently from any value that (after an eventual confirmation) might be attributed to the theory of relativity and gravitation."

When he is informed of the prize, Albert is also asked not to mention relativity in his official acceptance speech. This is an unnecessary recommendation, since he is unable to go to Stockholm. When presenting the prize on December 10, 1922, at the Nobel ceremony, the ever-present Svante Arrhenius emphasizes that relativity "pertains essentially to epistemology and has therefore been the subject of lively debate in philosophical circles."

The prize is then physically handed to him the following year but not in Stockholm. Albert receives it on July 11, 1923, in the context of the Nordic Assembly of Naturalists during the Gothenburg Tercentennial Jubilee Exposition in Sweden. Sitting in the first row, among others, is a spectator "eager to learn something about relativity": King Gustav V of Sweden.

ABSENT LAUREATES

Einstein's Nobel affair is certainly one of the most convoluted in the history of the prize. Looking back today, it seems inconceivable that the Academy took twelve years from the first nomination to actually award Einstein the prize and even more inconceivable that he was not awarded the prize for his work on general relativity.

Original studies and documents available from the archive of the Academy of Sciences lead to hypotheses that the main cause might have been the real difficulty on the part of Gullstrand and other members of the Academy to understand Einstein's work. We have already discussed how, during the first decades of the prize, Swedish academics strongly prioritized experimental results.

Some scholars have claimed political elements influenced the decision. The Swedish physics community had very close relations with the German one and might have been influenced by the rising hostility in Germany against Einstein and in particular against the theory of relativity. One cannot exclude the idea that at a certain point the German physicist's visibility in the media and popularity might have weighed negatively against him. This would be an external pressure to which the austere Academy of Sciences and in particular figures such as Gullstrand would be averse to, as if to prove their autonomy and imperviousness to any social pressure.

There is no doubt, however, that the event proves the impact of personal issues and individual beliefs, in certain cases, on the judgments expressed in the scientific arena. Without the obstinate hostility of Gullstrand, it is probable that Einstein would have been rewarded years earlier (or that he would have been rewarded for relativity). Without Oseen's commitment and the stubbornness, Einstein might never have been rewarded or might have been rewarded at a much later time.

Einstein's Nobel diploma is the only one in the history of the prize to contain a printed "warning notice" on the diploma handed to the scientist and in which the Swedish academics distanced themselves from a theory that was still considered controversial.

Nevertheless, this is not the only case in which a laureate is not present at the ceremony: it happens again in 2016 with the assignment of

 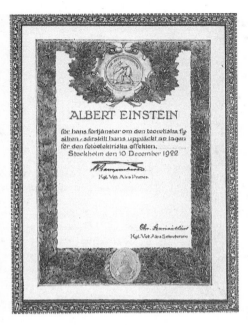

4.1 The original Nobel diploma for Einstein, including the "warning notice." In his presentation speech on December 10, 1922 (Einstein was absent), the Swedish physicist Arrhenius declared that relativity "pertains essentially to epistemology and has therefore been the subject of lively debate in philosophical circles." Courtesy of the Royal Swedish Academy of Sciences.

the Nobel Prize for literature to the American singer and songwriter Bob Dylan. His decision not to attend the ceremony causes much controversy.

However, there are other occasions in which the prize laureates are not able to attend the ceremony or do not wish to be present. Most cases concern literature and peace prizes (as mentioned earlier, at the very first ceremony, the writer Prudhomme was absent and asked the ambassador to pick up his prize in his place): among these are Ernest Hemingway, Boris Pasternak, Aleksandr Solzhenitsyn, Lech Wałęsa, and Andrei Sakharov. In the sciences, the most sensational case is probably that of the German scientists Richard Kuhn and Adolf Friedrich Johann Butenandt who with Gerhard Domagk refused their prize in 1939 due to the Nazi regime's boycott of the prize. That same year, due to the war, the prize to the only non-German—the American Ernest Orlando Lawrence,

13.

Inkom den 26.12.1921

Wykeham House; Oxford, Dec. 18/1921

Dear Sirs,
            Although the suggestions of a zoologist for the
Nobel Prize in Physics cannot carry much weight, yet even a
zoologist cannot fail to be stirred by the vibrations of thought
from another science when they are sufficiently strong and pene-
trating. Such seem to me to emanate from the work of Einstein
on Relativity, and I suggest his name as that of one who has stirred
the whole intellectual world.

        I am
            with much respect,
                Yours very sincerely,

                    Edward B. Poulton.

4.2 A couple of nominations that were sent to the Nobel committee for physics of the Academy of Sciences proposing Einstein for the Nobel Prize. Einstein was nominated more than sixty times between 1910 and 1922. The two nominations below both refer explicitly to relativity. The one on the left is from a prominent Oxford zoologist! Nobel Archives, vol. 1922, Stockholm. Courtesy of the Royal Swedish Academy of Sciences.

11.

Inkom den 31.10.1921

An das Nobelkomitee für Physik.

Den Nobelpreis für Physik beantrage ich für <u>Albert Einstein</u> Professor der Physik in Berlin.

Die von Albert Einstein im Jahre 1915 aufgestellte allgemeine Relativitätstheorie kann man als eine der grössten Thaten des menschlichen Geistes bezeichnen, sie hat befruchtend auf Astronomie, theoretische Physik und Philosophie gewirkt. Welche Anregung insbesondere die Astronomie aus ihr schöpft geht u.A. daraus hervor, dass bereits ein eigenes Institut, unter Leitung von Freundlich, besteht, das sich zur Hauptaufgabe gestellt hat die Folgerungen der Relativitätstheorie, z.B. die "Rotverschiebung", zu prüfen und für andere Probleme zu verwerthen.

                  Dr. B.Naunyn
                Prof.em. Un. Strassburg,
18.10.1921.   Baden B.

4.2 (continued)

4.3 Copy of the cheque for the prize intended for Einstein. Courtesy of the Harvard University Archives, HUM 132 (box 57, folder 7).

4.4 Albert Einstein receives the Nobel Prize in Gothenburg on July 11, 1923. Photo: Anders Wilhelm Karnell / Gothenburg Library Archive.

receives the Nobel Prize for physics "for the invention and development of the cyclotron"—is remarkably delivered at the University of California, Berkeley, in February 1940, with the university dean standing in for the king of Sweden and the medal being handed over via the Swedish consul in San Francisco. The same thing happens with the 1944 prize to Isidor Rabi, awarded at Columbia University in 1945.

In other cases, prize winners were not able to attend the ceremony because they were prisoners of war. Two examples are the German chemist Otto Hahn (prisoner of the English when he receives the prize in 1945) and the Austrian doctor Robert Bárány (awarded the prize for medicine in 1914, he was a prisoner in Russia and was able to make his speech in Stockholm only in September 1916). Lawrence Bragg, when he was awarded the prize for physics in 1915, was fighting on the front. Others, though not prisoners or at war, refused to take part in the ceremony to avoid meeting colleagues from enemy countries. Among these were William Bragg and Theodore Richards during the years immediately after the First World War.

Other laureates were not able to attend for health reasons. Among these was the Nobel Prize laureate for physics in 1962, Lev Landau.

One particularly dramatic example was the case of the Nobel Prize for medicine awarded to Ralph Steinman in 2011. It had been impossible to reach him to give him the good news, until it emerged that he had already died (it was decided on this occasion to maintain the prize although the Statutes of the Nobel Foundation do not allow the assigning of posthumous prizes; see chapters 1 and 2).

"NOBEL PRIZES WITH A BANG" AND THE LATE PRIZES: WILL MANY POTENTIAL WINNERS DIE BEFORE RECEIVING THE PRIZE?

The seventeen years between the publication of Einstein's article on the photoelectric effect (1905) and the assigning of the Nobel Prize (1922) do not constitute a unique case much less a record in the history of the prize.

Alfred Nobel's original will establishing the prize referred to discoveries and inventions that had taken place "during the previous year." Over time, this instruction was interpreted with increasing flexibility.

In 2010, the prize for physiology or medicine was awarded to Robert Edwards for "the development of in vitro fertilization" (IVF), a technique that had been tested since the 1940s with the first experiments by gynecologist John Rock and laboratory technician Miriam Menkin and subsequently developed by the same Edwards and Patrick Steptoe. In 1978, they were able to announce the news, acclaimed by doctors across the world, of the birth of Louise Brown, the first baby in history to be conceived in vitro. Since then, as is reported by the official communication of the Nobel committee of the Karolinska Institute, this technique has enabled the birth of over four million babies. If the birth of Louise Brown is taken as a reference point, thirty-two years went by before the decision was reached in 2010 to award the prize to IVF and its "fathers." Such a long time went by that Steptoe—deceased in the meantime—was not able to receive the prize (since it could not be awarded posthumously or in memory of): thus Edwards is the only one mentioned as the winner in physiology or medicine in 2010.

In some cases, the wait lasted up to forty-two years, as we saw in the case of Jack Kilby (rewarded in 2000 for his invention of the integrated circuit, which dated back to 1958; see chapter 2) and even fifty years—or fifty-three, to be precise. That is how long it took the Academy of Sciences to decide to reward the German physicist Ernst Ruska in 1986 "for his fundamental work in electron optics, and for the design of the first electron microscope," which dated back to 1933.

According to Carlo Rubbia, Nobel for physics in 1984 with Simon van der Meer, there are two ways of winning the Nobel:

The Nobel Prizes "with a bang," that are assigned straight after the discovery, within a minimum of one year, to allow the dust to settle. And those proclaimed after a long time. Ours was a Nobel with a bang, one of the rare cases of immediate consecration. The discovery, by myself and Van der Meer, of field particles W and Z was long overdue.[7]

Recently, the "Nobel Prizes with a bang"—that is, assigned shortly after the publication of the rewarded result—have become more and more rare, while the number of those awarded significantly later have increased.

This tendency struck a group of physicists who analyzed the data of the "delays" between the result and the prize between 1900 and 2013 and reached three fundamental conclusions.[8]

First of all, the delay increased for all three scientific subjects of the prize: physics, chemistry, and medicine. Until 1940, 61 percent of the Nobel Prizes in physics, 48 percent of those in chemistry, and 45 percent of the prizes in physiology or medicine were assigned within ten years of the discovery. Only 11 percent of the physics prizes, 15 percent of those in chemistry, and 24 percent of those in physiology or medicine were assigned twenty or more years after the discovery. From 1985 onward, only 15 percent of the prizes in physics have been assigned within ten years, and 61 percent after twenty years or more. There is a similar tendency in chemistry: 18 percent of the prizes were awarded within ten years, and more than half (50 percent) after twenty years or more. In medicine, since 1985, there have been fewer than one out of ten rewarded within ten years, and one out of two was rewarded after twenty or more years.

Second, partly because of this increasing delay in assigning the prize, the age of the laureates is significantly rising. If this tendency is projected onto current life expectancy, the conclusion by the authors of the study is drastic: if this trend continues, then by the end of this century, in the fields of physics and chemistry, the Nobel laureates' age at discovery would become higher than the life expectancy—that is, several potential prize winners will die before they receive the Nobel Prize.[9]

Third, in some fields and in physics, in particular, the tendency to procrastinate the award is even more obvious than in others.

How can these tendencies be explained?

First of all, we obviously have to take into account a few structural changes within the world of research and more generally in society, starting with the increasing number of research activities and number of active scientists. When the prize was founded, the relevant scientific communities were still relatively limited, while today there are estimated to be six million active researchers worldwide and more than three million scientific articles published each year in more than 46,000 international peer-reviewed scientific journals.[10]

According to Karl Grandin, director of the Center for History of Science at the Royal Swedish Academy of Sciences, the main problem is that there is an "overabundance of demand compared to the supply." In other words, there is an increasing pool of potentially merit-worthy results—a pool so abundant to have now created a reserve "whereby the

Nobel committees could do without any new nominations and still have enough discoveries to reward for the next ten years."[11]

Another relevant element is the caution with which the different Nobel committees have chosen to make their moves, especially after the first decades. There has been a rather flexible interpretation of Alfred Nobel's last wishes from this point of view, with a preference for awaiting the confirmation and sedimentation of the results within the respective subject matters. To wait for a long time before rewarding a discovery has been considered a lesser evil compared to the risk of rushing to award results subsequently called into question—a fact painfully experienced on a few occasions with the Nobel Prize in medicine (see later in this chapter). In some cases, new discoveries or results have led to the consolidating or reevaluating of results and hypotheses dating back a few decades. This was the case for Peter Higgs, awarded the prize in 2013 following the experimental confirmation provided by the ATLAS and CMS experiments at the CERN of his discovery from five years earlier "of a mechanism that contributed to our understanding of mass of subatomic particles."

The authors of the above cited study, however, think that there are other more substantial reasons. In their view, the tendency to defer the prize and to reward increasingly elderly scientists reflects, particularly in physics, a decreasing "frequency of groundbreaking discoveries in fundamental science" and more generally "the increasing objective difficulty in achieving progress."[12]

Similar considerations lead to more general questions that go well beyond the scope of this book. But there is no doubt that this is yet another important reason for looking into the most famous and significant scientific prize in history. The Nobel can be used, among other things, as a thermometer and indicator of the state and prospects of a subject and, more generally speaking, of scientific research.

LOBOTOMIES AND OTHER VAIN HOPES: PRIZES THAT WOULD HAVE BEEN BETTER NOT TO AWARD?

As we saw to some extent also in the case of Einstein, the choices by the awarding institutions have often been very cautious. For this reason, there have been relatively more cases that were assigned late or even "excellent" exclusions than prizes assigned too quickly or too rashly.

Nevertheless, some choices come across as being rather embarrassing today. In physics, the choice in 1912 to reward the Swedish inventor Nils Gustaf Dalén "for his invention of automatic regulators for use in conjunction with gas accumulators for illuminating lighthouses and buoys" was a singular one (for a year in which there were other candidates, such as Albert Einstein, Henri Poincaré, and Max Planck). Another invention of little significance turned out to be the "method of reproducing colours photographically based on the phenomenon of interference" for which the Franco-Luxembourgeois scientist and inventor Gabriel Lippmann won the prize in 1908.

Historians and commentators have considered some decisions for medicine "embarrassing" and even "disconcerting." The assigning of the Nobel Prize in medicine to the Danish Niels Finsen in 1903 for "his contribution to the treatment of diseases, especially *lupus vulgaris*, with concentrated light radiation, whereby he has opened a new avenue for medical science" was rather optimistic, to say the least. Finsen used ultraviolet radiation against various pathologies, including tuberculosis, smallpox, and even some forms of cancer. The success among his patients was undoubtedly true, but many of his colleagues lamented the absence of reliable experimental data and a proper comprehension of the processes used for the therapy. Later on, the successes of pharmacological therapies lead to Finsen's work being almost completely forgotten (moreover, the scientist passed away barely a year after he received the prize). The major social relevance of tuberculosis in those days may have contributed to influencing the choices of the committee toward Finsen and his methods.

The most unfortunate year for the Nobel Prize in medicine is possibly 1927. There are two prize recipients, whose weight in hindsight does not even add up to one. The Danish Johannes Fibiger is rewarded (with the prize set aside the previous year) for the "discovery of *Spiroptera carcinoma*," a parasite that was believed to cause cancer, a hypothesis that was later refuted and the prize defined as "one of the biggest blunders of the Karolinska Institutet."[13]

The prize in 1927 went to the Austrian psychiatrist Julius Wagner-Jauregg "for his discovery of the therapeutic value of malaria inoculation in the treatment of *dementia paralytica.*" Yes, you have read this correctly: inoculating malaria against paralytic dementia. But one has to go back to the context of those times. During those years, it was understood that some forms of paralysis and dementia were linked to syphilis, and in some cases, it was observed that a high fever could favorably influence the state of the patient. Wagner-Jauregg tried to cause fevers in his patients in various ways: by injecting them with streptococcus and staphylococcus cultures, with tuberculin, and finally by provoking malarial fevers. In 1921, out of the two hundred treated cases, fifty were able to return to work. Many other colleagues started experimenting with his method. He had many critics within the Nobel committee: professor Gadelius refused to support the decision, because "a doctor who injects malaria into a paralytic is a criminal." The Wagner-Jauregg method continued to be used nevertheless in desperate clinical cases until the beginning of the 1950s.[14]

The doubts on the choices made that year emerged immediately in the daily press: the *Corriere della Sera* wrote of the discovery by Fibiger that "the applications of these notions to the pathology of cancer are very rare and those to its therapy are, it is safe to say, nonexistent." Of Wagner-Jauregg, it wrote that "opinions are still conflicting on the definitive value of the method of this cure."[15] In this case, too, it is probable that a significant role was played by the social relevance of this pathology, from which patients had little hope of recovery and a high risk of stigmatization.

The most controversial prize for medicine of all times is most definitely the one awarded to the Portuguese neurologist Egas Moniz in 1949 "for his discovery of the therapeutic value of leucotomy in certain psychoses." Never heard of it? Does the concept of lobotomy ring a bell perhaps? This was the name under which the operation presented by Moniz in 1935 became famous and that consisted of a series of incisions to destroy the connexions between the prefrontal region and other parts of the brain—a rather drastic method against schizophrenia that Moniz tests (or rather has tested since he is confined to a wheelchair after a schizophrenic patient shot him in the leg) initially on twenty cases, then on another eighteen, having developed a series of dedicated surgical instruments, concluding that "prefrontal leucotomy is a simple operation, always safe, which may prove to be an effective surgical treatment in certain cases of mental disorder."

Moniz's technique spreads rapidly, especially in the United States, where the neurologist Walter Freeman and the neurosurgeon James Watts develop their own version. At the time the Nobel is given to Moniz, ten thousand leucotomy operations already were performed in the United States. The numerous negative side effects and the expansion of pharmacological therapies contribute to the decline of this surgical technique, especially after the mid-1950s.

In 2004, the daughter of one of the victims of lobotomy officially requested that the Nobel Foundation take back the prize from Moniz. The Foundation, not surprisingly, refused but declared it was "relieved that the medical profession can today offer much more humane and effective therapies for the severely mentally ill patients."

According to historians and commentators, the spread of the technique and the decision to award Moniz should be understood in the context of an alarming increase in hospitalization for psychiatric reasons (after the Second World War, almost half of public hospital beds were occupied by these types of patients) and very few alternative therapies. Furthermore, at the time, Moniz was a well-known and respected name, particularly for his introduction of pioneering diagnostic techniques such as the cerebral angiogram but also for his activity as a politician and diplomat (among other things, he was also the Portuguese foreign minister).[16] Moniz was not present at the ceremony, and his prize was therefore picked up by a delegate from the Portuguese embassy. On November 15–16, the *Corriere della Sera* dedicated a long article to the news of the Nobel to Moniz. The headline is spine chilling nowadays: "Patients of Frontal Lobotomy Happier and Less Intelligent."

That similar discoveries and techniques could have been awarded a Nobel Prize seems as unbelievable nowadays as the tortuous and convoluted saga of the Nobel to Einstein. These examples help us understand how the perception of originality, relevance, and the beneficial character—or not—of scientific discoveries and results are linked to the context and sensitivity of different periods in history. This appears to be all the more obvious for medicine, where the perception of the relation between risks and benefits and even "ethicality" of certain experiments has drastically changed over time.

Generally speaking, there is no doubt that some of the above-mentioned scientists could have easily been left out of the story of the Nobel Prize. In the next chapter, we talk about some scientists who could definitely have been included in the history of the prize but were left out.

5

HOW NOT TO WIN A NOBEL PRIZE: THE STORY OF LISE AND OTHER PRIZE GHOSTS

Rien ne manque à sa gloire, il manquait à la nôtre. [His glory lacks nothing; ours lacked him.]
—Inscription under the bust of Molière at the Académie Française, which never included him among its members

THE STORY OF LISE MEITNER, WHO WENT TO SWEDEN YET NEVER RECEIVED THE PRIZE

When they tell you that science has nothing to do with politics, remind them of this story.

It begins with a young woman who arrives in Berlin from Vienna in 1907. In Berlin, she meets a colleague, Otto Hahn. They work well together. She is a physicist, and he a chemist: in many of their collaborations, he is the muscle, and she the brain. Otto is excellent and very precise in carrying out experiments; she is formidable at interpreting them. They work side by side all day long, singing songs but always keeping a professional distance. In the evening, Otto goes home to his wife, and Lise to her rented room. Their work goes well, and Lise obtains important recognitions and responsibilities from the Kaiser Wilhelm Institut in Berlin. Lise is the only woman in a 1920 institute group photo taken during a visit by the Danish physicist Niels Bohr. In another photo, one of the few showing her at leisure, holding a cigarette, Lise looks more like a tormented writer than a scientist.

Lise is of Jewish origin, but even after Hitler's rise to power, she keeps her job at the institute thanks to her Austrian citizenship. However, when Austria is annexed to Germany in 1938, Lise becomes a German citizen and is subjected to the racial laws. She turns to Otto for assistance, to no avail.

Now in her sixties, Lise escapes from Berlin in July 1938 with only the clothes she is wearing, ten marks, and a diamond ring she could sell if necessary, a departing gift from Otto. She has with her neither a passport nor her precious laboratory notebooks. On the day of her escape, she works all day at the laboratory so as not to raise suspicions, then goes to Otto's home and from there to the station. She is tempted to turn back once she reaches the platform, but her Dutch colleague Dirk Coster awaits her on the train to help her enter the Netherlands. From there, with the help of Niels Bohr, she travels to Sweden, where the physicist Manne Siegbahn is setting up a new laboratory.

The decisive events that lead to one of the most consequential scientific discoveries in history—and yet not to a Nobel Prize—take place between 1938 and 1939. From Stockholm, Lise stays in contact with Otto, who was continuing "their" experiments with his colleague Fritz Strassmann. By bombarding uranium with beams of slow neutrons, Otto and Fritz hope to obtain a new heavier element, but Lise is not convinced. Otto and Lise even meet in secret on November 13 in Copenhagen. Lise sends Otto back to the laboratory with new instructions. On December 19, Otto writes to Lise again:

There is something so peculiar about the "radium isotopes" that for now we are telling only you. [. . .] Perhaps you can come up with some sort of fantastic explanation.[1]

Hahn and Strassman keep finding traces of barium, which goes against all expectations. At the last minute, on the basis of Lise's answer, Otto and Fritz add a paragraph to the draft of their article.

It is almost Christmas. Lise is invited by some friends to spend the holiday in a Swedish tourist destination. She is joined by her nephew Robert Frisch, a promising young physicist who is part of Niels Bohr's entourage. The two go out on the snow, he on his cross-country skis, she on foot.

Lise tells her nephew about Hahn's experiment and her doubts. Lise keeps thinking about a lecture by Einstein that she heard in Salzburg thirty years earlier. Einstein gave a brief explanation of his theory of relativity and its implications, one of which was that "to every radiation must be attributed an inert mass."[2] Together, Lise and Robert find an explanation of Hahn and Strassmann's enigmatic results. Following Bohr's example of considering the nucleus a drop of liquid instead of a rigid structure, they understand that Hahn split the uranium atom into two parts (barium), thus provoking a release of energy. The process reminds Robert of bacteria splitting. When he returns to Copenhagen, he asks a biologist colleague what the most appropriate English term would be to describe it. This is how the term *fission* is born.

Robert naturally tells his director about it. Bohr barely lets him finish his sentence before jumping out of his seat: "We have been such idiots, all of us! It's fantastic! That must be it!" Then he throws Robert out of his office, telling him: quickly, run and publish this important discovery as fast as you can!

The article by Hahn and Strassmann is published in the journal *Naturwissenschaften* on January 6, 1939. The two certainly could not have published it with Lise, a Jew, and furthermore one who had illegally fled from Germany. But the article does not include even a line of thanks. Lise and Robert send their own article to *Nature*. But even then, luck is not on her side. Bohr, who encourages them to publish as quickly as possible, leaves for two months to the United States. During the crossing, he cannot resist the temptation of talking about the extraordinary discovery. Thus, when the article of Lise Meitner and Robert Frisch ("Disintegration of Uranium by Neutrons: A New Type of Nuclear Reaction") with the first comprehensive explanation of the phenomenon is finally published in *Nature* on February 11, 1939, the surprise effect has worn off. By then, nuclear fission is a topic of conversation among physicists across the world. "The original insight, so difficult to arrive at, was taken to be almost intuitively obvious—once understood."[3]

Otto writes to Lise that the discovery is the "miracle" that can save his laboratory from increasing pressure from his colleagues who are closer to the Nazi regime. But from that moment, he does his best to exclude her. He describes the results as being purely chemical, with nothing to do with physics, and fuels the impression among his colleagues that these important results were achieved after Lise left Berlin.

The nominations for the Nobel Prize start pouring in. To be precise, they started arriving many years earlier. Lise was nominated for physics beginning in 1937 and for chemistry beginning in 1924, proof that her name and her work were very well known. She was often nominated by heavyweights such as Heisenberg, Bohr, and Planck (who nominated her every year from 1929 to 1936). James Franck, a Nobel laureate who left Germany in protest against the purge of Jewish and dissident scientists, described the discovery as "the most important in physics of the last ten years" and proposed a joint prize for Hahn and Meitner. The committees for chemistry and physics continue to pass the hot potato from one another. The war does not help matters. Between 1940 and 1942, prizes are not awarded. The tragic end of the Second World War highlights the importance of the discovery of fission but occasions doubts within the Academy of Sciences when it becomes obvious that a lot of the work on fission took place in the shadow of military secrets. Influential members of the committee such as Arne Westgren and Theodor Svedberg continue the push to award the prize to Hahn. Knowing that the Nazi regime would forbid him from traveling to Stockholm to receive the prize in 1944, they set it aside for 1945. Someone uses that year to advocate again to add Lise to the discovery. But Svedberg strikes a line through her name and writes "Other scientists." During the meeting, Göran Liljestrand, who was neither a physicist nor a chemist but a doctor, argues that Hahn, a German, should be immediately rewarded to prove that the Nobel committee would not be influenced by the Americans. Therefore, in 1945 Hahn is awarded the 1944 prize.

But fate is not favorable to him either. At the end of March, two chemists appear at Hahn's laboratory in Tailfingen, asking him to accompany them. Otto ends up as a prisoner at Farm Hall in England with nine other scientists suspected of participating in the German atomic program. Throughout his detention, Otto continues to insist that Lise was not involved in the discovery. After the tragic events of Hiroshima and Nagasaki, in fact, the press give a lot of visibility to Lise as "the Jewish mother of the bomb," but in a memorandum signed on August 8, Otto repeats that "Professor Meitner had left Berlin six months before the discovery" and that "while she was in Germany, the fission of uranium had been out of the question."[4]

On November 16, 1945, the English newspaper the *Daily Telegraph* announces the Nobel Prize for Otto Hahn. No one, least of all the Academy of Sciences, knows exactly where Otto is. Detained at Farm Hall, Hahn is not able to go to Stockholm for the prize until the following year. Lise continues to receive nominations for the prize in chemistry until 1948 and for physics until 1965, three years before her death. But she is never awarded the prize. In 1948, Otto Hahn is the one to nominate her.

Why did Lise Meitner not receive the Nobel Prize? We can point to a collection of reasons.

First, her work was at the crossroads of chemistry and physics and at the center of a competition between the two respective groups in the Academy. Chemists such as Svedberg were adamant that nuclear research belonged to chemistry, excluding physicists like Meitner.

Second, her approach to research conflicted with the relationships, ambitions, and personal goals of a few influential Swedish scientists such as Manne Siegbahn. He had unwillingly welcomed Meitner in his institute. Their research approaches could not have been more dissimilar: Siegbahn concentrated on constructing instruments and running experiments; Lise, on the other hand, was focused on planning and interpreting experiments. "She can't do anything with her hands," Siegbahn claimed. Siegbahn feared that a Nobel Prize for Meitner would raise her profile above his own, foreclosing his ambition of becoming the reference for nuclear research in Sweden. He therefore used his influence to hinder the candidacy of the Austrian scientist. In addition, Meitner refused the offer to take part in the American atomic program in Los Alamos in 1943 and made statements against the use of nuclear research for military goals after Hiroshima. Such public positions could also have been embarrassing for those like Siegbahn, who had maintained close links to government circles and financing.

Additionally, according to some scholars, the Nobel committee may have had incomplete information due to the political situation and may have attributed a decisive role to Bohr in the discovery of fission because of it.

Finally, the wider political dynamics weighed on the decision. Claiming fission as an entirely German discovery soothed the wounded nationalism of the German scientists, who maintained close relations with their Swedish colleagues at least for the first years of the war. Even after 1945, many Swedish scientists—and other colleagues from the allied countries—considered it important to support the reconstruction of German science after Nazism. In this perspective, Hahn appeared to be one of the most reliable scientists to support: although he was very patriotic, he had never taken sides with the Nazis. His visit to Sweden in 1943 had made a good impression, and his report from Farm Hall described him as a "very sensible man decidedly well disposed towards England and America." Hahn became a respected public figure: "the good German who had not been a Nazi, the pure scientist who had never worked on the bomb."[5]

When the prizes were announced in 1946, Lise understood that the game was over for her. She wrote a note to Hahn: "The chance that I might become your Nobel colleague is finally settled. If you are interested, I could tell you something about it."[6] There is no trace of any reply

5.1 Lise Meitner with her colleagues invited to a symposium in honor of Niels Bohr, Berlin, 1920. Six of the scientists in the photograph would go on to receive a Nobel Prize: Bohr (1922), Hertz (1925), Franck (1925), Stern (1943), Hevesy (1943), and Hahn (1944). Photo: Professor Wilhelm Westfall, courtesy of AIP Emilio Segre Visual Archives.

from Hahn. Who knows if during those days Lise ever picked up the photograph taken on a happy day in 1920 in which she and Otto are smiling next to Bohr. Six of her colleagues depicted on that photo had won the Nobel Prize in the meantime.

In 1955, Lise told James Franck that the missed Nobel was not an "open wound" and that in any case she would not have wanted it without sharing it with Frisch. She lived her last years in Cambridge, England. She received many honorary prizes during her lifetime and was posthumously awarded the honor of having a chemical element named after her: 109, baptized "Meitnerium."

Otto and Lise, two colleagues who hummed Brahms's music in their shared laboratory before the politics of science and the Nobel Prize came between them, died within two months of each other in 1968.

FRONT-PAGE FAKE NOBEL NEWS: THE STRANGE CASE OF THE PRIZES FOR EDISON AND TESLA, WIDELY PUBLICIZED BUT NEVER GRANTED

On November 14, 1909, a *New York Times* headline read "Nobel Prize for Edison." The subtitle elaborated that Thomas Alva Edison's name "is "the Only One Mentioned in Chemistry Award." The article cautioned that rumors were "particularly uncertain . . . this year" but insisted that Edison's was the only name in circulation. However, the prize was instead awarded to Marconi and Braun.

Two years later, the paper again trumpeted "Nobel Prize for Edison," though the subtitle more conservatively claimed that the prize for physics would "likely" be awarded to the American inventor. The article cited information gathered in Stockholm, then explained the prize for the reader's benefit, ending with references to previous recipients. However, Edison did not receive the prize that year either. The official announcement on November 7 assigned the prize for physics to the German Wilhelm Wien "for his discoveries regarding the laws governing the radiation of heat"—laws that carry his name to this day.

Following the announcement, the paper reported that "[Edison] regards [the prize] [. . .] as a reward for poor inventors." According to the article, at the annual reunion of the Ericsson Memorial Society of Swedish

Engineers, a long-time colleague of Edison, Edward H. Johnson, had declared that Edison

would refuse the Nobel prize [. . .] if it was offered him on the ground that it was Mr. Nobel's idea that the prize was to be awarded to a man who did not have sufficient means to carry his inventions to a possible conclusion and make them profitable to the world.

The prize, therefore, was not suitable for a "commercial engineer who puts into practice the things he invents."

And yet four years later, on November 6, 1915, the front page of the *New York Times* announced the news received from the "Danish correspondent of the *Daily Telegraph* in London":

The Swedish Government [?] has decided to distribute the Nobel prizes next week, as follows: "Physics—Thomas A. Edison and Nikola Tesla; Literature—Romain Rolland, French, Henrik Pontoppidan and Troels Lund, Danish, and Verner von Heidenstam, Swedish; Chemistry, Professor Theodor Svedberg."

The following day, the *New York Times* returned to the topic, with an article titled "Tesla's Discovery Wins Nobel Prize" based on an interview with the physicist and inventor Nikola Tesla. Tesla claimed not to have any official confirmation but seemed to accept the paper's source and went as far as to speculate on the motivation of the prize, declaring that

I have concluded that the honor has been conferred upon me in acknowledgment of a discovery announced a short time ago which concerns the transmission of electrical energy without wires.

Tesla spent the rest of the interview imaginatively describing future applications of the discovery:

For instance, wireless telephony would be brought to a perfection undreamed of [. . .] we will deprive the oceans of its terrors by illuminating the sky, thus avoiding collisions at sea and other disasters caused by darkness [. . .] ultimately all battles, if they should come, will be waged by electrical waves instead of explosives.

When requested to comment on the (alleged) prize for Edison, Tesla got off in a cavalier way—or maybe mockingly, who knows?—"Mr. Edison was worth a dozen prizes." As for Edison, he refused to comment when contacted by a reporter on his return from Panama.

When the official announcement was made a few days later in Stockholm, it appeared that this had all been a colossal mistake. Only Romain Rolland had been accurately identified as a Nobel recipient, for the literature prize. The 1915 Nobel Prize for chemistry went to the German Richard Martin Willstätter "for his researches on plant pigments, especially chlorophyll." The prize for physics went to Sir William Henry Bragg and his son, William Lawrence Bragg (as discussed in chapters 2 and 3).

Erroneous information persisted in the press for months after the ceremony and still appears in popular memory from time to time.[7] Were Edison and Tesla simply so famous and familiar to the press that it seemed natural to associate their names with the already famous prize in the hope therefore of having it accepted at face value and of perpetuating the news? Had someone leaked Henry Osborn's 1915 nomination of Edison as a confirmed award?

Indeed, it would have made sense for the Academy of Sciences to award prizes to Edison and Tesla. A significant faction of engineers supported a fairly literal interpretation of Alfred Nobel's last wishes. Did the will not mention "invention" as well as "discovery" and "benefit to humanity" and therefore practical benefits? In their view, the Nobel committees for physics and chemistry were snubbing this aspect.

Another group, however, felt that the prize assigned in Sweden by Swedish institutions, per Nobel's instructions, had been too dismissive of Sweden's own scientists and inventors. In 1912, a proper "revolt" took place. Applied engineers and scientists turned down the Nobel committee's proposal to recognize Dutch physicist Heike Kamerlingh Onnes and voted instead for the Swedish inventor Nils Gustaf Dalén, who had just lost his sight during an experiment.

In a year in which there were candidates such as Albert Einstein, Henri Poincaré, Max Planck, and the Italian physicist Augusto Righi, the Nobel Prize for physics went to Dalén "for his invention of automatic regulators for use in conjunction with gas accumulators for illuminating lighthouses and buoys."

"SLEEPING SICKNESS" AND THE MOST NOMINATED ITALIAN FOR THE NOBEL

At the beginning of 1902, at London's Royal Society, one of the oldest and prestigious scientific academies, news arrives of a so-called sleeping sickness in Uganda, an English protectorate at the time. Known today as African trypanosomiasis, the illness triggers fevers, pains, and finally coma. Since "it proved impossible at that time to find a senior and experienced investigator willing to go to Uganda,"[8] three novice doctors are sent at first, Aldo Castellani among them. Born in Florence in 1874, Castellani transferred to the London School of Hygiene and Tropical Medicine the previous year. He and two colleagues set up a small laboratory in Entebbe, Uganda. On October 14, one of the three doctors, G. Carmichael Low, writes to the Royal Society:

I have the honour to inform you with great satisfaction that Dr. Castellani has found a germ to be the cause of sleeping sickness. As the credit of this discovery is certainly due to Dr. Castellani.[9]

Low attaches a preliminary note from Castellani and expresses his hope for its publications. Castellani asks to prolong his mission for another few months in order to spend more time investigating his observations on the pathogenic agent that at the time he thought to be a streptococcus. The Royal Society commission responds with a telegram saying "interesting results. You can stay another six months. Please send us your preparations and cultures immediately."[10]

But at this point, the commission is already on high alert. At the beginning of the following year, once the "seriousness and the 'imperial' importance of the illness" is recognized, the Royal Society decides to send Colonel David Bruce, a pathologist in the English army with extensive experience with infectious diseases, to Uganda. Castellani fears that Bruce's arrival might complicate the attribution of the discovery and insists that his preliminary note be published, but the Royal Society refuses. Castellani continues developing his experiments and realizes that it is not a *Streptococcus* that is responsible for the illness but rather a *Trypanosoma*. Castellani returns to London in April 1903. His article "On the Discovery of a Species of Trypanosome in the Cerebro-Spinal Fluid of Cases of Sleeping Sickness" is read by the Royal Society on May 14, 1903, and subsequently published in the *Proceedings of the Royal Society* and then picked up by the medical journal *The Lancet*. The discovery and the results of the commission on sleeping sickness even attract the attention of the daily press and of the London *Times* in particular.

Castellani will continue to claim authorship of the discovery. The Royal Society will insist that Bruce played a significant role and claim that Castellani was able to develop the definitive interpretation of his results only after his arrival. Despite the controversy, Castellani will go on to become the founder of the bacteriology laboratory of Ceylon, a professor in the United States, an author of other important discoveries in the field of infective illnesses, and the private doctor of Pope Pio XII, Mussolini, and Marconi. He was made senator for life in Italy and baronet in the United Kingdom, among other honors. His role during the war in Ethiopia was considered fundamental for the defense of the health of the Italian troops. Faithful to the monarchy even after the advent of the Republic, he followed the royal family to Lisbon, where he died in 1971.

Castellani also holds a high rank in an unfortunate list: that of the scientists most frequently nominated for a Nobel Prize who nevertheless never received one. Castellani received sixty-one nominations over the course of nearly fifty years.

The record for unsuccessful nominations goes to a scientist whom you have probably never heard of, although our health owes him a lot: the French veterinarian and biologist Gaston Ramon, who developed the diphtheria vaccine. Ramon was nominated for the prize in physiology and medicine 155 times over more than twenty years between 1930 and 1953. Next on the list, with 115 nominations, is the doctor Émile Roux, also French and the author (with Louis Pasteur) of important discoveries on diphtheria and the rabies vaccine. According to Erling Norrby, permanent secretary of the Royal Swedish Academy of Sciences between 1997 and 2003, the candidacies of Roux and Ramon might have been penalized by the fact that some discoveries had already won prizes in the fields of immunology and infective illnesses—for example, the Nobel to the German physiologist Emil von Behring in 1901 for his work on serums and in particular on the antidiphtheria serum. The same reason could have led to Castellani's numerous nominations having been ignored, although in his case one cannot exclude the possibility that his key role in Italian medicine during the Fascist era might have penalized him.

Though a high number of nominations is significant, other factors may discount their value. In the case of the French René Leriche, most of the nominations came from his countrymen, with little recognition from non-French scientists. In other cases, the nominations seemed to be merely formal, from one colleague to another, and thus did not convincingly reflect the merits of the proposed scientists.[11]

Table 5.1 The ten most nominated scientists who never received the Nobel Prize

	Nominations	Period	Subject
Gaston Ramon	155	1930–1953	Medicine
Émile Roux	115	1901–1930	Medicine
Arnold Sommerfeld	84	1917–1951	Physics
René Leriche	79	1930–1953	Medicine
Jacques Loeb	78	1901–1924	Medicine
Albert Calmette	77	1907–1934	Medicine
Rudolf Weigl	75	1930–1939	Medicine
Christophe Ingold	68	1940–1965	Chemistry
Walter Reppe	63	1949–1962	Chemistry
Aldo Castellani	61	1905–1953	Medicine

"DROWNED" BY THE PRIZE, OR THE NOBEL PRIZE FOR CHEMISTRY ACCORDING TO PRIMO LEVI

There are the so-called inert gases in the air we breathe. They bear curious Greek names of erudite derivation which mean "the New," "the Hidden," "the Inactive," and "the Alien." They are indeed so inert, so satisfied with their condition, that they do not interfere in any chemical reaction, do not combine with any other element, and for precisely this reason have gone undetected for centuries. As late as 1962 a diligent chemist after long and ingenious efforts succeeded in forcing the Alien (xenon) to combine fleetingly with extremely avid and lively fluorine, and the feat seemed so extraordinary that he was given a Nobel prize.[12]

This is the beginning of *The Periodic Table* (1975), in which chemist and writer Primo Levi recounts stories associated with chemical elements. The discovery that Levi refers to is the one by the British scientist Neil Bartlett. Bartlett overturned a chemistry dogma by proving—thanks to an experiment on xenon, the "heaviest" of all the noble gases—that the so-called noble gases were not inert. The result was published in an extremely brief article (less than 250 words) that appeared almost immediately in June 1962 in the journal *Proceedings of the Chemical Society*. But, contrary to what Levi wrote, Bartlett never received the Nobel Prize. His discovery stayed impressed in the memory of his colleagues to the point that many chemists, including Primo Levi, remained convinced that Bartlett did receive the prize.

The list of scientists who could have received one of the scientific Nobel Prizes is long. From the French mathematician and physicist Henri Poincaré (he received fifty-one nominations between 1904 and 1912) to the abovementioned Canadian doctor Oswald Avery (nominated thirty-eight times over more than twenty years) and the father of psychoanalysis, Sigmund Freud (nominated thirty-two times for physiology or medicine and once for literature), the names of some of the greatest scientists of the twentieth century are missing from the prize's roll of honor.

The "missed prizes" can be roughly divided into two categories.[13] The first includes scientists whose discoveries or results were never rewarded in Stockholm. The second includes the so-called fourth persons: scientists such as Lise Meitner who were excluded from the list of prize winners although the recognition went to a discovery they had taken part in. The term "fourth person" refers to the rule whereby no more than three individuals can be rewarded per year and has been informally extended to cover any instance in which a scientist who participated in the work is not recognized, whether they were excluded because of the cap on honorees or overlooked for another reason.

Bartlett, Castellani, and Avery are included in the first category. Their colleagues—particularly those from the Karolinska—were slow to recognize the consequences of Avery's discovery that DNA transmits genetic information. Avery himself, a shy and even-tempered character, was very cautious about his results. He was

a poor salesman for his own discovery, fastidious and hypercautious, almost neurotically reluctant to claim that DNA was genes and genes were simply DNA.[14]

Other scientists are included in this category, some of them well known to the public, such as the American doctors Jonas Salk and Albert Bruce Sabin, both remembered for having developed two distinct poliomyelitis vaccines and both ignored by the Nobel committee. This decision was possibly influenced by the conviction that this field of discovery had already been rewarded in 1954 with the Nobel to John Enders, Frederick Robbins, and Thomas Weller for having cultivated poliovirus cells in tissues. Or perhaps they were penalized by the violent controversy that raged between the two scientists and their respective vaccinations.

Other multinominated scientists—such as the German physicist Arnold Sommerfeld and other brilliant scientists of the twentieth century such as John Desmond Bernal, who is considered nowadays one of the pioneers of molecular biology, or Leó Szilárd, the first physicist to imagine how a nuclear chain reaction could actually work—were possibly penalized by the fact that they had produced high-quality work on numerous themes but had not associated their names to one single discovery in a decisive way.

"THE MOST IMPORTANT PHOTO IN HISTORY" THAT DID NOT WIN THE NOBEL (OR DID IT?)

We wish to suggest a structure for the salt of deoxyribose nucleic acid (D.N.A.). This structure has novel features which are of considerable biological interest. [. . .] This structure has two helical chains each coiled round the same axis. [. . .] It has not escaped our notice that the specific pairing we have postulated immediately suggests a possible copying mechanism for the genetic material.

With these words, laced with caution and understatement, the young scientists from the Cavendish Laboratory of Cambridge, James Watson and Francis Crick, describe their discovery of the structure of DNA in their article "Molecular Structure of Nucleic Acids: A Structure for Deoxyribose Nucleic Acid," published in the journal *Nature* on April 25, 1953.

The article was published (without any peer review—that is, without any revision from colleagues, which today would be unthinkable) a few weeks after the discovery that took place during the first days of April 1953. Such a short article (one page) and such a small image had never had such a huge impact on science, society, and culture.

The famous image of the double helix—"an authentic icon in natural sciences," according to art historian Horst Bredekamp,[15] that changed the representation of life transmission from then on—was designed by Odile Speed, the wife of Francis Crick, a clothes designer and painter of nudes, based on a sketch by her husband.

There are no other images other than the drawing in Watson and Crick's article. However, there is one photograph in one of the two other articles published in the same issue of *Nature*. It is not just any photograph but the famous "Photo 51," the mother of all intuitions on DNA structure, that made Watson jump as soon as he saw it:

> The instant I saw the picture my mouth fell open and my pulse began to race. The pattern was unbelievably simpler than those obtained previously ("A" form).[16]

This photo is at the center of the dispute over the paternity of one of the most important discoveries in the history of human sciences—and one of today's most famous and obvious Nobel Prizes (although as we will see, this was not the case at the time).

It is an image of DNA obtained by X-ray diffraction in May 1952 by a doctoral student, Raymond Gosling. Gosling worked with Maurice Wilkins and Rosalind Franklin at King's College, but he showed the photo without Franklin's authorization to Watson, who found in the photo one of the decisive insights for their discovery.

The article by Watson and Crick briefly cites Wilkins and Franklin. Thanks to his friendship with the coeditor of *Nature*, the director of the King's College laboratory was able to obtain also the publication of Wilkins's article in the same issue. At that point, at the insistence of the author, Franklin's article with the photo was also included. In his autobiography *The Double Helix*, Watson's version of his encounter with the image is a little different:

> To my surprise, he [Maurice Wilkins] revealed that with the help of his assistant Wilson he had quietly been duplicating some of Rosy's [Rosalind Francklin] and Gosling's X-ray work. Thus there need not need a large time gap before Maurice research efforts were in full swing. Then the even more important cat was let out of the bag: since the middle of the summer Rosy had had evidence for a new three-dimensional form of DNA. It occurred when the DNA molecules were surrounded by a large amount of water. When I asked what the pattern was like, Maurice went to the adjacent room to pick up a print of the new form the called "B" structure. [. . .] Moreover, the black cross of reflections which dominated the picture could arise only from a helical structure.[17]

Photo 51 has inspired numerous debates, television programs, and even a theater play, mostly centered on the nonrecognition of Rosalind Franklin as the coauthor of the discovery. According to the BBC, it could be "the most important photograph ever taken." It was Wilkins, during the summer of 1953, who built the first accurate DNA model based on Watson and

Crick's intuition and to verify it with data like those obtained by photo 51. "Of course the structure was right—commented the molecular biologist Brian Sutton a few years ago—it was too beautiful not to be."[18]

Watson and Crick felt the Nobel Prize was in their pocket by then, but they actually had to wait a few years before they could celebrate in Stockholm. The reactions from their colleagues were, in fact, rather cautious. For example, a year later, the German chemist and Nobel Prize laureate Otto Warburg had not even heard of their work. Crick, who did not have a doctorate yet at the time of the discovery, did not even obtain a permanent position at the Medical Research Council. Notwithstanding a diffused "journalistic mythology," even the press was initially hesitant. The first newspaper article appeared on May 12, 1953, quite a few weeks after the article in *Nature* by Watson and Crick. It presented the discovery as "an isolated result, obtained by anonymous scientists in an unknown laboratory."[19]

Even the two bold protagonists started having some doubts: "Jim and I were worried we had made been badly mistaken and that we would cover ourselves with ridicule once again," Crick said years later. It took years for their colleagues and in particular the members of the Nobel committees to grasp the importance and implications of their discovery. The first explicit reference to the value of the discovery appeared in 1957 in a document by the Nobel committee for chemistry. Citing a colleague, it mentioned that the Watson and Crick model "was beyond the state of conjecture." That same year, a nomination from John H. Northrop (Nobel Prize for chemistry, 1946) contributed to encouraging the Swedish academics to understand the importance of what was happening:

Nucleic Acids, then, are the stuff of life, for which so many men have searched for so long. This, I believe, will come to be considered as one of the greatest of all chemical and biological discoveries.[20]

The Nobel committee members realized they had made an egregious error in overlooking Oswald Avery—the scientist of Canadian origin who had been the first one to demonstrate that genetic information was transmitted by DNA, not by protein molecules as believed until then—for a prize. But by then it was too late: Avery died in 1955 after having been unsuccessfully nominated thirty-eight times over twenty years. Later, the Swedish chemist Arne Tiselius, one of the most influential members of the Nobel committee, admitted that the omission of Avery had been one of the most sensational oversights in the history of the prize.

The prize for medicine to Watson and Crick, initially nominated for chemistry, was considered an extraordinary opportunity to make up for the omission (see chapter 3). The role of Lawrence Bragg, *deus ex machina* of the Cavendish Institute, proved crucial. That year, Bragg successfully proposed two other scientists from Cambridge, Max Perutz and John Kendrew (nominated for physics but rewarded for chemistry), and a member of the rival team, Maurice Wilkins. But while there were no doubts about the candidacy of Watson and Crick, Wilkins was included, according to Erling Norrby, only "as a pawn to compensate the guilt of the Cavendish group towards King's College."[21]

Rosalind Franklin died of cancer in 1958 and was never considered for the Nobel Prize. Possibly annoyed by the posthumous reevaluation of Franklin's role, both Watson and Crick were critical of her character and personality. In Stockholm, during their official speeches at the Nobel Prize reception, Watson and Crick chose not to talk about DNA. The only one to hold a conference on the subject was Wilkins, making a brief reference to Franklin.

Ten years had gone by since that photograph had made Watson's heart beat.

"FOURTH PERSONS": THE ITALIAN WHO CAME CLOSE TO THE NOBEL PRIZE TWICE (ACTUALLY, MAYBE THREE TIMES)

Rosalind Franklin was not technically a "fourth person" when the prize was awarded to her colleagues, since she had passed away by the time the prize was awarded. Undoubtedly, her name could legitimately have appeared alongside that of the other laureates—instead of Wilkins, some say.

In 1902, the Nobel for medicine was awarded to the English military doctor Ronald Ross "for his work on malaria, by which he has shown how it enters the organism and thereby has laid the foundation for successful research on this disease and methods of combating it." The award ruffled feathers, because it ignored Italian zoologist Giovanni Battista Grassi's significant contribution to demonstrating the role of the *Anopheles* mosquito in transmitting the illness. Grassi and Ross had been involved in a harsh dispute, each one claiming priority for the discovery, with serious mutual accusations. Probably contributing to his exclusion, among other reasons, were his zoological approach centered on analyzing and classifying the parasite (compared to Ross's medical approach) and the influence of former Nobel laureate Robert Koch, with whom Grassi had had bitter discussions.[22]

The Italian physicist Giuseppe "Beppo" Occhialini was a "fourth person" on two occasions, although both times only one scientist was actually rewarded. Occhialini was nominated thirty-three times between 1936 and 1965. In 1932, Occhialini and the British physicist Patrick Blackett discovered the positron at almost the same time as the American Carl Anderson. The prize went to Anderson, who published his discovery a few weeks before they did. In 1948, the Academy of Sciences "made amends" by awarding the prize to Blackett, who in the meantime had become a very visible political figure thanks to his antimilitary positions during the war. In 1950, another British scientist, Cecil Powell, was awarded the prize for his discovery of the pi-meson (or pion), to which Occhialini and the Brazilian César Lattes had contributed.

Another important exclusion was most certainly that of Jocelyn Bell from the Nobel Prize for physics in 1974. Jocelyn Bell had been a student of Antony Hewish when she made some observations that led to the discovery of pulsars, for which Hewish and Martin Ryle received the 1974 prize in physics. Bell's exclusion elicited strong reactions from some of her colleagues. British physicist and astronomer Fred Hoyle openly accused Hewish of having stolen Bell's data and excoriated the Nobel committee in the press. According to some commentators, these interventions—and other unorthodox positions on various themes—might have contributed to Hoyle's own exclusion from the Nobel Prize in 1983, when the Indian scientist Subrahmanyan Chandrasekhar and the American William A. Fowler were rewarded, respectively, for the "theoretical studies of the physical processes of importance to the structure and evolution of the stars" and the "theoretical and experimental studies of the nuclear reactions of importance in the formation of the chemical elements in the universe." Hoyle and Fowler had, in fact, worked closely together and cosigned an important article on the subject in 1957.

Jocelyn Bell was invited to Stockholm to the Nobel Prize ceremony in 1993, when the discovery of a new type of pulsar was rewarded.

Anders Bárány, the secretary of the Nobel committee for physics at that time, handed her a small reproduction of the Nobel medal and told her, "Unfortunately I can't do more than this."

An American doctor of Armenian origin, Raymond Vahan Damadian, took his exclusion much less gracefully. In 2003, following the announcement of the prize to Paul Lauterbur and Sir Peter Mansfield "for their discoveries concerning magnetic resonance imaging," Damadian and some of his supporters paid for an advertisement titled "The Shameful Wrong That Must Be Righted" in American newspapers (the *New York Times*, the *Washington Post*, and the *Los Angeles Times*) and a Swedish newspaper (*Dagens Nyheter*), openly inviting the Swedish academics to reconsider their decision. The advertisement opened with a controversial image of the Nobel medal upside-down and ended with a prepaid coupon to be sent to the Nobel committee for medicine. Quoting patent documents and scientific publications, the ad claimed paternity of the discovery for Damadian. Some colleagues and commentators recognized his relevant contribution, some claimed that the attribution of the prize was legitimate, and others underlined how it depended on what was being rewarded—the visionary capacity of imagining the solution (attributable to Damadian) or its finalizing from a practical point of view. One comment in the *New York Times* pragmatically put an end to the debate by underlining colorfully the irreversible nature of Nobel decisions: "No Nobel Prize for whining."

Similar situations underline the contrast between the prize's legacy of focusing, on the one hand, on individual merits and, on the other, on research and innovation processes that, particularly in certain fields, were increasingly becoming a collective process that included heterogeneous networks—groups of researchers, technicians, companies, institutions, and users.[23]

"THIRD PERSONS" OR THE RINGO STARR EFFECT

Just like the Nobel medal, the recognition of scientists has two faces.

The collective and organized dimensions that have increasingly characterized research, especially during the last ten years and in some fields, make it complicated to attribute individual paternity and merit and

5.2 Advertisement published in various international newspapers in response to Damadian's failure to receive the Nobel in October 2003.

therefore lead not only to exclusions (like the case of the "Fourth persons") but also to inclusions.

We could call it the "Ringo Starr effect," from the name of the drummer for the Beatles. Ringo Starr joined the Beatles later than the other three band members and often referred to his luck in finding himself at the right place at the right time:

I've never really done anything to create what has happened. It created itself. I'm here because it happened. But I didn't do anything to make it happen apart from saying "Yes."[24]

The honor roll of Nobel laureates includes scientists who found themselves with the right colleagues at the right time. They would probably never have been rewarded if the prize had been individual, but the possibility of rewarding three scientists for the same discovery enabled them to be included with their colleagues. Maurice Wilkins, awarded the medicine prize with James Watson and Francis Crick in 1962 for the discovery of the structure of DNA, was, according to many scholars, the classic example of a "third person." Another case that was different but had a similar outcome was that of Robert Curl Jr., who in 1996 shared the prize for chemistry with Harold Kroto and Richard Smalley for "their discovery of fullerenes." Some say that if it had been possible to reward four scientists, two other scholars would have been included, Huffman and Kratschmer, "who had produced fullerenes for the first time in measurable quantities." Since, however, it was impossible to reward four people or to reward many of the collaborators and doctoral students who took part in the study, the committee decided to reward three of the five authors of the decisive article published in *Nature*.[25] As Curl discovered, however, the media and the public were often inclined to "annul" this effect, thereby ignoring the "third person." From day one, the calls for interviews concerned only Kroto and Smalley, and his name was often left out when the discovery was referred to. Wilkins is also often forgotten by the public and the media: when one thinks about the discovery of the structure of DNA, one always thinks of the Watson-Crick couple (as one young scientist said when he met Crick: "I always thought it was a single person, Watson-Crick").

Paradoxically, some "fourth persons" or scientists who were never rewarded—such as Lise Meitner or Rosalind Franklin—have become part of the collective memory much more than some prize recipients. When Jocelyn Bell was asked if she had been saddened by not receiving the prize, she replied: "Why should I be sad? I made a career of not having the Nobel Prize."[26]

The destiny of "third persons" is to be unexpected beneficiaries and, at the same time, left at the margin of the Nobel glory—a little like Ringo Starr, the third person (actually fourth) who made history yet, at the same time, was always in the shadow of the talent and fame of John Lennon, Paul McCartney, and George Harrison.

ALL THE COLORS OF THE NOBEL PRIZE AND ITS MOST ELUSIVE GHOST

Among the ghosts of the Nobel, one of them possibly gets the top prize. No category can describe him properly. His name does not appear on the Nobel Prize list; in fact, almost no one today even remembers him. Yet his discovery obtained multiple recognitions in the history of the Nobel Prize, it continues to be taught to students, and we all benefit from it when we take a vitamin or read the label on a food product or a product composition.

Mikhail Semënovič Tswett[27] is born in 1872 in Asti to an Italian mother and a Russian father. His mother dies soon after his birth, and the young Mikhail is raised and educated in Switzerland. He is a bright child, graduates in botanic studies, and immediately wins the Davy Medal for work on the anatomy of the Solanaceae. He reunites with his father in Russia, determined to continue his studies, but there he discovers his Swiss degree is not recognized and that he will have to repeat his doctorate. In 1901, he transfers to Warsaw as a laboratory assistant. There, working on plant pigments, he discovers that by transferring a pigment solution obtained from leaf extracts into a column of calcium carbonate, the pigments are trapped in the top part of the column. By using a solvent, the pigments move at different heights based on their different absorption capacity, highlighted by different colors:

Like light rays in the spectrum, the different components of a pigment mixture, obeying a law, are separated on the calcium carbonate column and can thus be qualitatively and quantitatively determined. I call such a preparation a chromatogram and the corresponding method the chromatographic method.[28]

The name chosen by Tswett for his discovery, *chromatography*, refers to the analogy with light rays in a spectrum, with the different colored zones corresponding to the various vegetal pigments of the original experiment, and it probably also plays with the scientist's surname ("color" in Russian).

Tswett succeeds in demonstrating that the method can also be used for many other compounds and presents his discovery for the first time in March 1903 during a meeting of biologists in Warsaw. During the following years, he publishes it first in Russian and then in an important German botany journal. His doctoral thesis is published and wins an award. The press also shows interest in Tswett's discoveries, but his botanist colleagues are mostly indifferent to them. His academic career struggles to take off and instead turns into something of an ordeal. His application for a position as a botany professor in Moscow is turned down, and he is not keen to go to Siberia, where there would be a position. His health deteriorates, and he takes some time off in Odessa. During his absence, Warsaw is occupied by German troops, and he loses everything: personal belongings but especially manuscripts, books, and notes. After a few other unsuccessful attempts, in 1917 he finally is offered a position in Tartu as a professor of botany. But again fate turns against him, and the following year Tartu is also occupied by the Germans. Evacuated to Voronezh, Tswett dies a little while later from a bad throat inflammation on June 26, 1919.

In 1918, Tswett is nominated for the Nobel Prize in chemistry. However, the letter that recommends him for the prize, by the Dutch Cornelis van Wisselingh, mainly refers to "the research on chlorophyll and other pigments." The problem is that a Nobel already was awarded on this topic to the German Richard Willstätter in 1915. Furthermore, Willstätter is a fierce rival of Tswett, whose work he knows well and often criticizes, possibly because it is at times in competition with his. The 1918 prize for chemistry, one of the most controversial ones in the history of the Nobel, is awarded to Fritz Haber, the father of ammonia synthesis and the lethal gases used during the First World War.

It will take another ten years for the work by Tswett to be rediscovered and for it to become clear that this was not a discovery limited to botany but a formidable method for the analysis and separation of organic and inorganic substances. In 1952, the Nobel Prize for chemistry was awarded to Archer J. P. Martin and Richard L. M. Synge "for their invention of partition chromatography." A total of twenty-five scientists who were awarded the Nobel between 1937 and 1999 used chromatography techniques to obtain their results. Today its numerous variations are used in various techniques that are widely applied in the food industry, medicine, industry, and forensics to analyze compounds and isolate principles.

Perhaps unintentionally, the epitaph on Tswett's tomb captures the tragic destiny of this pioneer who tried in vain to talk to botanists about chemistry and to chemists about a chemistry for which they were not yet ready. He was always in the wrong place at the wrong time, yet he was part of the history of the prize and of science like an invisible thread while always in the shadows: "Mikhail Semënovič Tswett, May 14, 1872–June 26, 1919. He invented chromatography by separating molecules and uniting people."

6

DOES THE PRIZE MAKE THEM MORE APPEALING?
THE IMPORTANCE OF BEING A NOBEL

TUCO: Hey, amigo! You know you got a face beautiful enough to be worth $2,000?
BLONDIE: Yeah, but you don't look like the one who'll collect it.

—S. Leone, *The Good, the Bad, and the Ugly*

"THE WINNER TAKES IT ALL": JOYS AND SORROWS OF THE MATTHEW EFFECT

Timing is everything. In 1996, a committee of British experts turned down the funding request of their colleague Harold Kroto. Two hours later, the Royal Swedish Academy of Sciences announced the Nobel Prize for chemistry would go to Robert Curl Jr., Richard Smalley, and Harold Kroto "for changing the way we think in physics and chemistry with their discovery of fullerenes." The British committee had to backtrack and reverse its decision by giving Kroto the money.[1] Thanks to the recognition from Stockholm, the British chemist was admitted into the exclusive circle of so-called visible scientists: elite researchers whose public recognition accords them almost bulletproof prestige and a reputation that can open just about any door.

The founder of the sociology of sciences, Robert K. Merton, understood this phenomenon and its cumulative effect. He called it the "Matthew effect," from the Gospel passage that says "For unto every one that hath shall be given, and he shall have abundance; but from him that hath not, shall be taken away even that which he hath." (Matthew 25:29):[2]

Those who already have visibility and prestige will have privileged access to other resources and opportunities for visibility, and so on

[. . .] a scientific contribution will have greater visibility in the community of scientists when it is introduced by a scientist of high mark than when it is introduced by a scientist who has not yet made his mark.[3]

In the words of a Nobel laureate in physics, "the world tends to give credit to [already] famous people."[4]

Analyzing some empirical data, Merton and his students discovered that essays submitted to a scientific journal were more frequently accepted if a Nobel laureate or a particularly well-known researcher were among their authors. Similarly, essays from a scientist were more often cited by their colleagues after they had been awarded a widely known prize such as the Nobel.[5]

As a paradigmatic case, Merton recalls the story of Lord Rayleigh, Nobel laureate for physics in 1904. His name had been accidentally omitted from a manuscript presented to the British Association for the Advancement of Science. The committee turned it down, thinking it was "the work of one of those curious persons called paradoxers."[6] As soon as the real author was discovered, the manuscript was accepted. Merton considered these mechanisms to be due to the poor "recognition" capacity in science and the rigidity of its allocation system. In the illustrious Académie Française, where only forty places were available, the "forty-first chair" included the likes of René Descartes, Blaise Pascal, Jean-Jacques Rousseau, Denis Diderot, Stendhal, Gustave Flaubert, Émile Zola, and Marcel Proust.

Merton considered the Matthew effect to be "dysfunctional for the careers of single scientists, who are penalized during the initial stages of their activity," but functional for science in general, winnowing the huge quantity of results, publications, and other projects. Furthermore, the names of famous scientists were able to attract the community's attention to particularly innovative discoveries that would otherwise have struggled to be taken into consideration.

It is difficult—extremely difficult—to become a celebrity scientist. But once acquired, celebrity feeds on itself: "once a Nobel laureate, always a

Nobel laureate."[7] Thus, the Nobel Prize is often a prelude to further recognition and benefits.

The physicist Robert Millikan received twenty honorary university degrees and sixteen significant prizes after he was awarded a Nobel; the chemist Harold Urey calculated the financial benefits deriving from his Nobel as "four to five times the sum received for the Prize."[8] Many British laureates were later offered the title of baronet and commemorated on stamps. Three Italian laureates were nominated as senators for life: Guglielmo Marconi, Rita Levi-Montalcini, and Carlo Rubbia. According to Kary Mullis, Nobel Prize for physics in 1993, the prize is a sort of "universal access key":

Nobody in the world doesn't understand the weight of the Nobel Prize. Once you have it, there is not a single office in the world that you can't go into. If I call them and say, I would like to talk to you about something, and I'm so-and-so, the Nobel laureate, they'll see me at least once. It opens every door.[9]

Hiroshi Amano received approximately four hundred requests for conferences per year. After he was awarded the Nobel Prize for physics in 2014 for the invention of light-emitting diodes (LED), the number of invitations increased by 1,000 percent. The Nobel, he claimed, gave him the opportunity to explain the importance of his research for environmental protection and to accelerate its industrial applications.[10]

Some Nobel laureates have tried turning their notoriety to political ends. On January 14, 1992, 104 laureates signed a public appeal for peace in Croatia published by the *New York Times*. When Salvatore Luria (Nobel Prize for medicine 1969) received a telegram of congratulations from President Richard Nixon, he immediately replied with another telegram asking the president to end the American intervention in Vietnam.

Often Nobel recipients have suffered from their own popularity. "We are swamped by letters and visits from photographers and journalists," complained Marie Curie in 1903 after her first Nobel for physics (she would be one of the very few to receive a second one, for chemistry). Francis Crick (medicine, 1962, with James Watson, for the discovery of the structure of DNA) drafted a standard form to apologize for being "unable to accept your kind invitation to . . ." with check boxes for ". . . deliver a lecture . . . cure your disease . . . be interviewed . . . appear on TV . . . write a book . . . accept an honorary degree . . ."

From:
M.R.C., *Laboratory of Molecular Biology, Hills Road, Cambridge.*

Dr. F. H. C. Crick thanks you for your letter but regrets that he is unable to accept your kind invitation to:

send an autograph	read your manuscript
provide a photograph	deliver a lecture
cure your disease	attend a conference
be interviewed	act as chairman
talk on the radio	become an editor
appear on TV	contribute an article
speak after dinner	write a book
give a testimonial	accept an honorary degree
help you in your project	

6.1 Form sent by Francis Crick to decline invitations, ca. 1963. *Source:* Francis Crick archives, Wellcome Collection, London.

Einstein summarized his experience with his usual irony: "As punishment for my contempt for authority, destiny has made me an authority."[11]

SLAP SCIENCE (AND THE WIFE'S HAND) ON THE FRONT PAGE!

On the evening of November 8, 1895, the German physicist Wilhelm Conrad Röntgen is in his laboratory at the University of Würzburg conducting experiments on cathode rays. For some time, he has noticed that some rays are able to leave traces of the bodies they hit. Just over a month later, on December 22, 1895, he manages to take the first X-ray in history: of his wife's left hand, showing her bones and wedding ring. A new type of radiation has just been discovered.

Röntgen hurries to communicate this important result: he quickly writes an article of about ten pages titled "On a New Kind of Rays," which is accepted by the journal *Sitzungsberichte der Physikalisch-Medizininschen Gesellschaft zu Würzburg* on December 28, 1895. The article does not contain any photographs, but Röntgen includes some with copies of the article that he sends to some of his European colleagues on New Year's Day, 1896.

Röntgen's friend and colleague Franz Exner, on receiving the article and photos, showed them during a dinner at his home. Among Exner's guests was another physicist, Ernst Lecher. Lecher asked Exner if he could borrow the images for a few days and immediately showed them to his father, the editor of the Viennese newspaper *Neue Freie Presse*.

On January 5, 1896, the newspaper's front page announces "A Sensational Discovery," underlining the great contribution that the rays would make in the field of medicine for the diagnosis of illnesses. The photo of Anna Bertha's hand with a ring—which had reportedly caused Röntgen's astonished wife to exclaim "I saw my death!"—caused a particular stir.

Picked up immediately by an English correspondent from the *Chronicle* in Vienna, the news and image quickly reach the press on the other side of the Atlantic: within a few days, newspapers like the *New York Times* are talking about them. In the United Kingdom, the London *Standard* writes that far from being a prank, Röntgen's discovery opens a vast field in scientific research. By the end of January, *Nature* publishes an English translation of Röntgen's article, and at the beginning of February the *British Medical Journal* dedicates a long article to the discovery of this photography and its applications to medicine and surgery. The most hidden structures would soon be revealed.[12]

In Italy, the *Corriere della Sera* also mentions Röntgen's discovery on its front page on January 15, 1896:

We have conquered another eye, so to speak. Who can say what other spectacles we will be offered though the fixed visions of this new outlook, how many mysteries of nature's deep workings will be revealed, how it will enable us to clearly and simply understand phenomena that were until now reserved to a very few after long and difficult research?

Two days later, the paper returns to the subject after receiving many messages from its readers and underlines the interest and the enthusiasm that the X-rays are raising among scientists and the general public:

The physics professor [of the University of Padua] presented an extremely successful experiment tonight of a photograph of a hand with Röntgen's method. The professor was loudly applauded.

On April 29, after a few clarifications provided by Röntgen in a letter from Berlin, the *Corriere* is already able to offer an example of the possible applications of X-rays in an article titled "The Röntgen Rays on the Wounded in Africa":

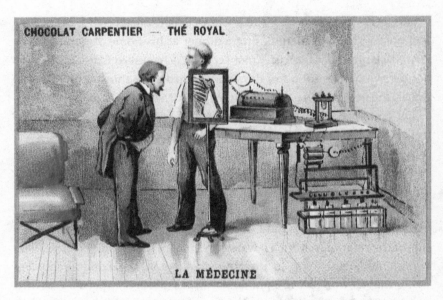

6.2 In very little time, Röntgen and X-rays come to symbolize the frontiers of medicine. Their popularity was such that they were used in advertising communications. In this chromolithograph, part of a series made for promotional purposes by a chocolate producer (1896–1900), the scientist is depicted while he observes the torso and arm of a man. Photo: Wellcome Collection, London.

The experiments with Röntgen's rays have continued on the wounded in Africa, enabling the extraction of a bullet situated near the nervous vascular fascia in the right arm of a soldier named Mussiano.

In a few weeks, the whole world has heard of the rays that Röntgen abbreviated "in all modesty," according to the *Corriere*, to *X*. Röntgen thus has become a celebrity. By the end of 1896, more than a thousand books and articles have been written about X-rays and their applications. Röntgen is bombarded with requests for interviews and visits to his laboratory.

In 1901, the assignment of the first Nobel Prize for physics, just instituted, seems almost obvious. Röntgen receives it "in recognition of the extraordinary services he has rendered by the discovery of the remarkable rays subsequently named after him." It is not completely without controversy, however. The German physicist Philipp von Lenard claims to have anticipated most of Röntgen's findings. He would go on to receive the prize for his work on cathode rays in 1905.

During the following years, the Nobel Prize for physics is repeatedly awarded for other research related to X-rays. In 1914, Max von Laue obtains the prize "for his discovery of the diffraction of Röntgen's X-rays by crystals"; the following year, the British scientists William H. Bragg and William L. Bragg receive it "for their services in the analysis of crystal structure by means of Röntgen's X-rays." In 1918, Charles G. Barkla receives the prize (reserved in 1917) "for his discovery of the characteristic Röntgen radiation of the elements," and in 1924 the Nobel Prize for physics goes to the Swede Manne Siegbahn "for his discoveries and research in the field of X-ray spectroscopy."

Röntgen has opened what seemed to be a never-ending source of discoveries and research. It therefore is noteworthy that by the time that he receives the award, Röntgen's personal interest in X-rays has already vanished. After the first historic article on X-rays, he writes one in 1896 and then another in the next year, and then he does not seem to care about them—to the point that, in a rather rare case in the history of the Nobel, he collects the prize in Stockholm but does not even give his own Nobel lecture about his discovery, promising to return at a later date but never does.

"THE WIRELESS MAN" AND BOUNDLESS CELEBRITY

On December 8, 1909, the Swedish paper *Tidningen Kalmar* reports that "Marconi is on his way to Stockholm to receive the Nobel Prize." The following day, another Swedish paper, *Svenska Dagbladet*, gives a detailed report of the Italian physicist and inventor's arrival. Two journalists wait for his train at the station in Stockholm and follow his transfer to the Grand Hotel. Marconi's stay in Sweden is followed minute by minute by the Swedish press, as might befit a Hollywood celebrity or a rock star. Journalists report on his clothing (a fur coat), his companions (his wife and her sister), and the parties and receptions they attend. There is the evening in an Italian restaurant, his meeting with the king, his visit to Uppsala, the dinner organized in his honor by the Italian ambassador, Bottaro-Costa. The papers publish photographs, drawings, and even cartoons of him.

During his stay in Stockholm, Marconi is repeatedly interviewed by the Swedish press. A first interview on his arrival at the Grand Hotel in Stockholm is published on December 9 by the *Dagens Nyheter* under the title "A Quarter of an Hour with Marconi." Another article in the *Aftonbladet* presents him as "the great wireless man" and ends by quoting "the long line of interviewers waiting outside" for their turn to ask him a few questions. One interviewer even goes so far as to ask for a portrait of Mrs. Marconi, to which the scientist laughs and answers, "You'll have to make do with mine, because I don't think we have one [of my wife]." Another journalist, hoping to please him, replies to a comment by Marconi on the characteristic darkness of Stockholm in December by declaring that the Nobel celebrations should be organized in May or June. When Marconi leaves at the end of his stay, the *Aftonbladet* writes that it hopes "we would soon have news from the wireless man, perhaps via his wireless invention."

Sweden's interest in Marconi and his results predated his Nobel Prize. Since 1897, the Swedish papers had dedicated many articles to the scientist, supported by photos, diagrams, and maps. The press was particularly captivated by the possible applications of Marconi's work to commerce and navigation, as well as the figure of the scientist himself, who had "no academic background."[13]

The Italian press and public opinion also were well acquainted with the figure and works of Marconi long before he was awarded the Nobel Prize. On July 4, 1897, the *Corriere della Sera* publishes on its front page the news of Marconi's invention (which the article calls his "discovery") with a long article that continues onto the following page and includes diagrams and detailed explanations. The article combines, in an interesting fashion, a technical style with a more imaginative style that approaches fiction:

We don't quite know what ether is: but is it really necessary to define it? Does the human soul need a definition in order to understand the minutest graduations of feelings and the wildest torments of passion?[14]

Toward the end, the article features a conversation between Marconi and an English journalist, apparently taken from the foreign press:

—So you could send a dispatch from this room to the whole of London?—asked the correspondent from the *Strand Magazine* to Marconi [. . .].
—Without any doubt! With the relative power instruments of course one could.
—Through all the houses?
—Through all of them.
The two speakers were then at Marconi's home in Westbourne Park, eight or nine kilometers from the central post office [. . .].

—I am actually currently working with Precce to establish regular communications between the English coast and a lighthouse ship. It will be the first practical application of my invention.

—Do you surely foresee others even more marvelous? [. . .]

Marconi smiled and stopped . . . looking at the journalist and almost embarrassed. But in that smile and in that look were reflected all the hopes and dreams of glory of the young inventor.[15]

Almost two years later, on April 23, 1899, the Sunday supplement of the *Domenica del Corriere* features on its back page a colored illustration of Marconi, explaining his "experiments of the wireless telegraph across the Channel" to a group of gentlemen. At the turn of the new century, Marconi is already a familiar figure to readers: the *Corriere della Sera* writes regular updates about his successes and the diffusion of his invention, trumpeting with great national pride his "triumph" over the initial skepticism expressed by the great American inventor Thomas Edison. Edison acknowledges Marconi's success by declaring to the American press that "Marconi has succeeded in casting an electrical spark to the other side of the Atlantic." The quote is immediately relayed by the Italian press.

On December 17, 1901, a long article is dedicated to the first wireless transmissions between England and North America. "I have given the world a magnificent Christmas present," declares a proud Marconi. On December 31, the *Corriere* reports Marconi's imminent marriage in New York to the "beautiful Miss Giuseppina Holman from Indianapolis. . . . The couple met in 1899 on board the *Saint-Paul* en route to Europe."[16]

The attention paid by the press to Marconi and his activities continues to grow during the following years. His visits to Europe—to Italy in particular—are met with enthusiasm, and he is offered a series of awards and honors, including honorary Roman citizenship and membership in numerous scientific academies and societies. "A vote of applause and recognition to Guglielmo Marconi, thanks to whom Italy's name is covered in glory" is mentioned in the minutes of the Chamber of Deputies (January 30, 1903). Journalists are sent to his country of origin to uncover colorful details of his adolescence and precocious talent. On a few rare occasions, there is also some criticism from abroad on a few technical issues—mainly due to the fact that Marconi's system may not be completely adequate for protecting message secrecy.

Figure 6.3 provides an overview of the articles published by *Corriere della Sera* on Marconi and his activities between 1901 and 1911: 215 articles in ten years—an impressive figure, given the relatively few pages in the newspaper at that time; 67 articles in a single year, 1903, were devoted to Marconi. That year, Marconi is featured on the cover page of the Sunday illustrated supplement *Domenica del Corriere* (January 4, 1903): the caption describes him as "The Hero of the Day" (figure 6.4).

Ironically, one of the years in which Marconi is less mentioned is the year in which he is awarded the Nobel Prize (1909): only four articles were written, and only three of them actually mention the prize. The first one very briefly gives the news of the prize in just six lines published on November 16:

The *Svenska Dagblatt* [sic] announces that the Nobel Prize for physics will be divided between Guglielmo Marconi, inventor of radiotelegraphy, and professor Carlo Ferdinando Braun of Strasburg, who carried out important research on the invention.

The news is confirmed a few days later in another short article that describes the other laureate for physics—mistakenly called "Bauer"—as the man "who had perfected the system of wireless telegraphy." Finally, a long article published on December 11, 1909, reports in more detail

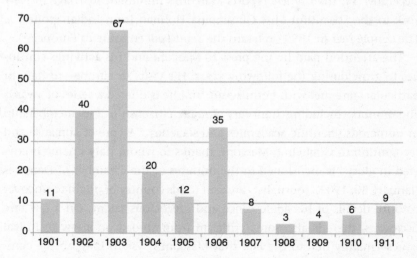

6.3 Number of articles on Guglielmo Marconi in the *Corriere della Sera*, 1901 to 1911 (*N*=215).

6.4 "The Hero of the Day": Marconi on the cover of the *Domenica del Corriere*, January 4, 1903.

on the prize ceremony that took place in Stockholm the previous day and includes brief biographies of the laureates and descriptions of their respective contributions. Possibly because he was already very well known to readers, Marconi is only briefly mentioned in the article. Braun's research is contextualized "in that wonderful field of electric waves that was explored by Marconi and had a fortunate influence on the various systems of wireless telegraphy."

The relevance and practical benefits of Marconi's invention quickly became obvious to the wider public, and his international reputation resonated with the emerging national pride of a newly unified Italy. His habitual visiting of royal families and heads of state was a source of excitement and curiosity that turned him into a sort of *ante litteram* member of the international jet set. Various studies define him as one of the most notable examples of the development of a "cult of celebrity" in the media that characterized the years between 1890 and 1910. With the development of a consumer society and the media industry, personalities from the world of culture, science, and entertainment were given increasing space to the detriment of entrepreneurs and politicians.[17]

The Nobel Prize seems to have had a modest impact as far as Marconi's visibility and popularity are concerned. When he received the prize in Stockholm, his invention had already been celebrated and was familiar to the Italian public, to the point of being old news. But in the years following the prize, Marconi's new projects continued to be mentioned in the media. On May 22, 1910, the *Corriere della Sera* announced "a happy event in the Marconi household":

Mrs. Beatrice Marconi, the wife of our illustrious countryman Guglielmo Marconi, who has been living for over a month in his villa in Pontecchio, has been delivered of a beautiful boy. The news was immediately sent to Marconi, who is now traveling from America to England.

HOW TO MAKE A PRIZE FAMOUS? BY (ALSO) REWARDING SCIENTISTS WHO ARE ALREADY FAMOUS

The case of Marconi is certainly the most emblematic example of prize celebrity during the first years of the award, but it is also seen that in the case of Röntgen, the name of the scientist and his discovery of X-rays were already amply familiar to the media and international public opinion.

Another relevant case is that of Marie Curie. When she received the prize for physics in 1903 with her husband, Pierre, "in recognition of the extraordinary services they have rendered by their joint researches on the radiation phenomena discovered by Professor Henri Becquerel," her name and that of the "new element" radium had already amply circulated for several months in the pages of newspapers, especially in England and Italy. A great stir was caused by the couple's invitation to the Royal Institution in London, Marie Curie being the first woman to be invited to the sessions and "the whole of London wanting to see up close the parents of radium. Professor and Madame Curie were invited to parties and banquets." And in November 1903, about a month before the announcement of the Nobel, widespread coverage is given to the awarding to the Curies of the Davy Medal from the Royal Society.[18] On April 18, 1903, the *Corriere della Sera* dedicates a long article to radium, defining it as

a scientific mystery studied and discussed in front of the general public with a breadth that would fill most readers in other countries with stupor. Not only has the great *Times* opened its welcoming columns to controversies on the atomic theory, even the working-class *Daily Mail* has interviewed top scientists on the matter.

On August 25, the newspaper printed another article ("Mysterious Forces. X-Rays and Radium"), claiming that "Radium has become a news item well known even to modest readers."

And even after the announcement of the Nobel, the Italian paper underlined how their work was already well known abroad, although it was not completely recognized by French institutions:[19]

the work of the Curries [sic], now awarded a prize in Scandinavia, has already been widely recognized for its importance by English scientific academics, while in France it has been so overlooked that the famous chemist has not been able to obtain a position at the University of Paris because he lacked certain titles. Bureaucracy could not have suffered a worse blow than the one inflicted on it by the Nobel Prize commission.[20]

Similar examples contribute to piecing together one possible answer to one of the questions that has always surrounded the Nobel Prize. What conferred on the Nobel Prize, in a relatively short time after its establishment, a visibility and prestige such that it became a sort of global brand that makes the names of Nobel laureates visible and recognizable, as they are today?

Without underestimating the importance of the amount of the prize, the notoriety of the founder, the quality of the choices, or the proverbial discretion during the selection process, there is no doubt that the prize was enhanced by the preexisting visibility of some of the early discoveries—X-rays, wireless telegraphy, radium—and the popularity of some of the early laureates, such as Wilhelm Röntgen, the Curies, and especially Guglielmo Marconi.

A NOBEL IS FOREVER: SCIENCE BETWEEN REPUTATION AND VISIBILITY

"Papa, our schoolmates say you are cool!" That is how Peter Agre's children welcomed their father when news got out that he had been awarded the Nobel Prize for chemistry in 2003. "Until then, they had never taken any interest in my work," commented the surprised scientist when he understood that he had crossed the line from being a science celebrity to a celebrity scientist.

In an insightful essay, sociologist Alessandro Pizzorno distinguishes three possible meanings of reputation: (a) "excellence in a role that a person must perform"; (b) "the capacity to be credible" (as in "situations of interpersonal relationships, and tends to coincide with the notion of trust"); and (c) "the capacity to be visible."[21]

By its nature, reputation in the sciences seems primarily of type (a). Each scientist is judged by peers, experts in the same field, who evaluate results, requests for financing, or written works proposed for publication.

However, numerous studies have underlined how some criteria linked to the personal credibility of researchers are often decisive for judging the validity of results and in particular the resolution of scientific controversies. In the controversy over the existence of gravitational waves, which developed in physics particularly in the late 1960s and early 1970s, uncertainty combined with experimental methods and results to lead to social criteria being taken into account.

These included the reputation of the experimenters and institutions, their nationalities, the level of inclusion in the main scientific circuits of the research subject, and informal information from collaborators and other colleagues. Once the reliability of the experiments and researchers had been established, it was easy to ascertain whether the gravitational waves existed.[22]

In many situations, the attribution of reputation among scientists seems to be similar to what Pizzorno describes as a *community* reputation: meaning a reputation centered on "his behavior as a member of the community and his capacity to incarnate its values and support its norms" rather than on the person's specific characteristics.[23] In the controversy over cold fusion, some of the most critical judges focused less on the results than on the behavior of the scientists who were thought to be violating the rules of their community. The decision to present the results during a press conference before having them published by a specialized journal was particularly stigmatized.[24]

From a certain point of view, the Matthew effect can be described as a process through which the judgment of a reputation for excellence becomes one of a reputation for credibility. The judgment on credibility then shapes and supports the reputation for excellence.

There is a level beyond which the reputation of some scientists turns into visibility, exceeding judgment by peers and exposing them to a public celebrity not dissimilar to what characterizes sports and entertainment personalities.

The most obvious example of scientific excellence of the last century is Albert Einstein. In November 1919, confirmation of the theory of general relativity became a front-page story for all the major newspapers worldwide, and after that every trip or lecture by the physicist became an event of public relevance. During his six weeks lecture tour in Japan (November–December 1922), he was acclaimed as a superstar, with thousands paying to attend his lectures and trains being forced to make additional stops so that more people could welcome him at every station. He appeared on the cover of *Time* magazine four times. The image of him sticking his tongue out, from a photograph taken in his car on his seventy-second birthday, is still reproduced on posters and T-shirts today.

According to Goodell, the phenomenon of "visible scientists" emerged particularly in the second half of the twentieth century, having to do with both the degree and quality of their fame. Some scientists become increasingly visible not just by virtue of an "internal" reputation that reflects on their public one (such as the case of the initial Nobel prizes) but also for their ability to respond to—and exploit—the operational logics of the mass media:

> Circumventing the traditional channels for influencing science policy, some scientists take their message directly to the public. To succeed, they must be knowledgeable, articulate, dramatic, persistent, and sophisticated about press operations. Those who succeed become known to the public not only for their science but for their public involvement.[25]

Numerous studies have shown that the criteria used by the media to select "scientific experts" for opinion or comment do not necessarily align with those of the scientific community. Among the criteria used by journalists in their selection of these experts is visibility outside the research community (possibly due to advisory roles or popularization activities), accessibility in relation to the resources and limited time available to journalists, ability to be interesting from a human point of view, availability to comment on a broad range of questions, and possibility of easily justifying their choice (as members of a particularly prestigious institution or recipients of prestigious prizes and acknowledgments).[26]

Visibility in the media tends to follow a pyramid structure similar to the ones identified by Merton and other scholars in the scientific community. At the top of the pyramid are an extremely limited number of celebrities who are consulted on themes that are often outside their sphere of competence (and often outside the field of science), while a wide base of scientists is consulted much more sporadically. Visibility is thus subject to a recursive effect of the type identified by Merton with the Matthew effect, where previous visibility and media presence tend to translate into greater visibility and media presence.

In combining scientific prestige and visibility within the community, the Nobel Prizes represent perhaps the epitome of how a reputation inside the scientific world and public visibility can, under certain circumstances, condense, self-reproduce, overlap, and mutually reinforce each other. The confirmation of this overlapping comes paradoxically from cases such as the "only announced" Nobel Prizes to Edison and Tesla: these were so well-known to the public and the media that it was taken for granted that they should receive the most important prize.[27]

In recent decades, the communication dynamics triggered by an award such as the Nobel appear to be even more intense. On one hand, the proliferation of content and the competition for attention means that relying on information selection heuristics such as the Matthew effect is all the more inevitable. The number of active scholarly peer-reviewed journals has been estimated to be over 46,000, and the number of papers published yearly over three million.[28] On the other hand, the phenomenon of scientific celebrities should be understood in the framework of a growing "mediatization" of science that has made the world of research more permeable to the mechanisms of visibility and public communication.[29]

In a similar scenario, Nobel laureates become precious expendable communication resources and not just in the sciences. They offer the media the possibility of personalizing themes and complex and abstract topics, while linking the most disparate subjects to the prestige of the prize and thus ennobling them: "Levi-Montalcini: Fewer Cars, More Buses" (headline of the *Corriere della Sera*, November 9, 1986). The scientist was interviewed on the problem of traffic in Rome.

This use of the Nobel Prize as a benchmark for authority, a rhetorical resource, and a source of prestige to legitimize certain positions—in the scientific field, but not only—is particularly notable in the media. A substantial portion of the articles citing Nobel Prize laureates in the daily press (even one out of two of those published by the *Corriere del Sera* and over four out of ten articles in the English newspaper *The Times*) do it in this manner:[30]

Also Nobel Prize laureate Renato Dulbecco, father of the genome, approves the joint initiative by to the American president and the English prime minister on the Human Genome Project. (*Corriere della Sera*, March 16, 2000)

Yes to euthanasia from two Nobel Prize laureates. (*Corriere della Sera*, December 6, 2000)

In my view, the new Central Charities Commission of the Cariplo Foundation is of very high profile. Two Nobel Prizes. . . . (*Corriere della Sera*, January 28, 2001)

The prize was awarded by the president of the jury, James Black, Nobel Prize for medicine in 1988. (*Corriere della Sera*, February 20, 2001)

The Politecnico di Milano is an institution that had among its ranks the Nobel Prize for chemistry Giulio Natta. (*Corriere della Sera*, June 29, 2001)

Xenotransplantation is not a novelty, but when the news is confirmed by Nobel Prize recipient Renato Dulbecco, it bolsters the prospect of it becoming a reality in the near future. (*Corriere della Sera*, September 8, 2001)

More than a hundred frontline researchers of the first rank in the field of medicine, including five Nobel laureates, have launched an attack against excessive regulation in the field of research that, they claim, enables rivals in other countries to overtake them. (*The Times*, June 13, 2000)

letter signed by 110 of our country's most prominent researchers, among whom are five Nobel laureates and 38 fellows of the Royal Society. (*The Times*, June 13, 2000)

The celebrity of the Nobel laureates, like that of other figures from the world of entertainment known to the general public, also reflects on social initiatives and events:

The recipient of the Nobel Prize for medicine [Rita Levi-Montalcini] sat in the front row with her usual black velvet dress. (*Corriere della Sera*, October 8, 2001)

Two grand ladies were greatly admired. The Nobel Prize recipient Rita Levi-Montalcini (extremely elegant in sparkling black) and the ballerina Carla Fracci (in light white trousers). (*Corriere della Sera*, March 29, 2001)

In the same way, the presence of a Nobel Prize laureate on the board of a tech company can become a precious instrument for legitimization. A visit to a research institute or support for a fundraiser can have significant effects on the visibility of institutions or initiatives. It can even make history in the world of entertainment, as was the case at the end of the 1990s, when Renato Dulbecco was invited to co-host the Sanremo Festival.

PUT AN(OTHER) NOBEL IN SANREMO

Who knows if Alfred Nobel, who took up residence in Sanremo during the last years of his life, would have ever imagined a curious event that had one of "his" laureates as its protagonist roughly a century later.

On February 23, 1999, the forty-ninth Sanremo Festival began. The festival is the most popular song competition in Italy, broadcasted live during prime time by Public National Radio and Television since 1951. Three conductors hosted that year's edition: professional TV presenter Fabio Fazio, international model Laetitia Casta, and a scientist: Renato Dulbecco, Nobel Prize laureate for medicine in 1975. He explains his reasons for accepting the host's invitation as follows: "I thought a scientist on the stage of the festival would be a good way to promote science."

During his public performance at the festival, Dulbecco actually speaks very little about science in general and even less about his own research activity. In his many press interviews, the questions mainly focus on his experience living in the United States, his personal relationships with other famous personalities from the world of science and culture such as Nobel laureate Rita Levi-Montalcini (whom he confesses to having had a crush on as a young man), his taste in music, and how he is preparing for his experience in Sanremo. He never explicitly talks about the contribution that won him the Nobel Prize, which is referred to vaguely as "a discovery that contributed to the polio vaccine." The articles highlight his youthful demeanor despite his advancing age, his brilliance, and his sense of humor.

Over the evenings of the festival, Dulbecco is limited mostly to presenting some of the singers in the competition and exchanging a few words with the other presenters. Only during the last evening does he briefly mention his hopes for the future progress of cancer research and the need to support it, announcing that he wishes to donate his fee for his participation in the festival to that field of research.

Dulbecco's presence and role during the festival move between two popular narratives of the scientist and the role of the scientist in society.

According to the first narrative, the scientist is "something different": a genius, someone who knows things that we know nothing of and who has seen things that we cannot even grasp. According to the other narrative, the scientist, even though he is a Nobel Prize laureate, still remains "one like us." In this case, the image of science and its protagonists that is emphasized is more reassuring and close to everyday life. This is not about a scientist who manipulates genes or builds uncontrollable machines but rather about one who works humbly and patiently day in and day out, who comes across as an attainable model for anyone with the determination to commit to hard work and research.

In fact, the three presenters assume, from the point of view of roles, a clear triangular arrangement: on the sides, two exponents of perfectly symmetrical and equally coveted and valued qualities in our society—the beauty of Laetitia Casta, who came from the glittering world of top models, and the intelligence of Renato Dulbecco, the Nobel Prize laureate who until then had been investigating the hidden secrets of nature. Despite their very different professions, the two have this "specialty" in common—being different from the public who watches them, underscored by the elegant dress worn by one and the particularly formal attire of the other. Two stars that were seemly light years apart from us common mortals.

In between them, the real host of the show, Fabio Fazio, inevitably assumes the role of "common mortal," a spokesperson for the dreams and expectations of the public with whom he goes to great lengths to identify. Dressed less formally than the other two, he makes a show of deference to the extraordinary people standing next to him.

"You were fourteen, the age at which my friends and I were trying to learn how to drive a scooter, when you invented the electronic seismograph. How did that idea pop into your head?" Fazio asks Dulbecco with exaggerated astonishment.

The difference between the host and the Nobel laureate is also emphasized by the different attitudes that the two have toward the French model. While Fazio plays the stereotypical common man enthralled by the star's inaccessible beauty, Dulbecco makes a show of being relatively indifferent, a man with much more important things on his mind. "Laetitia Casta is a slightly complicated large sponge," is his curt judgment, a phrase that is extensively commented on by the media. In this sense, the presence of Dulbecco signifies a descent among the "common people" of an otherwise unattainable personality. During *TG1* (the television evening newscast on the public Rai 1 channel) on February 22, 1999, Fazio declares that

A man who has made substantial contributions to research on, let's say, the polio vaccine, and who decides at one point in his life, at the age of 85, to take a five-day holiday, is giving an incredible lesson to all of us.

At the same time, despite his "specialness" and detachment from the mundane aspects of life that characterize the scientist's public image, during the festival it is emphasized that Dulbecco still remains a normal person, "one of us." "He is like us," claims Fazio repeatedly during his conversations with the press. And Dulbecco reinforces this interpretation of his participation in the festival by declaring on *TG1* on February 22:

I am aware that some people say—why would a scientist come here? Well, I mean, first of all, *a scientist is like everyone else, and I am here to a certain extent to prove that my interests are the same as all of yours and that deep down there is nothing extraordinary about being a scientist.* [italics added]

He did, however, confirm a few of his peculiarities as a scientist during the *TG1* interview of February 24:

—So during the first evening, were you nervous like the rest of us?
—Actually, as far as I'm concerned, seeing the public or talking to the public isn't a novelty because I often hold conferences. . . .

Some studies on scientific discourse identify the main source of humor in science and among scientists as being the short circuit between opposite discursive repertoires. Thus, a humorous effect is produced when the "special" character of science is combined with events that characterize our daily life or when the informal register used by scientists in "private" contexts, such as the laboratory, contrasts with the more formal tone of conferences and scientific papers.[31]

This is particularly visible during Dulbecco's participation in the Sanremo Festival, especially during the satirical news program *Striscia la Notizia* (Strip the News). One *Striscia* incursion is based on the scientific theme of cloning, which had been getting media attention after the announcement of the birth of the first cloned animal, Dolly the sheep, a few years before. An actor dressed as Dulbecco and explicitly introduced as his "clone" circulates among the participants and the guests of the festival and succeeds in deceiving them.[32] The stunt takes on a double symbolism. On one hand, it rebels against science, based on a classic narrative model with a scientific theme of particular relevance and public visibility: cloning. On the other hand, it brings Dulbecco's "special character" back down to earth: he really is one of us in the sense that anyone can replace him, if not in his area of expertise at least in his public persona. The satire works because this is the aspect of the scientific persona that counts in the public's eye and the world of television. No one speaks of Dulbecco's discoveries. No one is interested in them. The only message he conveys is that of being present and representing science, not himself:

And here, among real flowers that seem fake (and vice versa), the scientist Dulbecco says of Laetitia Casta that she is a large complicated sponge when the girl doesn't look anything like a sponge and does not appear to be overly complicated. Thus, science comes across as being slightly ridiculous, within everybody's reach. (*Corriere della Sera*, February 27, 1999)

The short circuit between discursive repertoires is complete when at the end of the festival the same satirical news program shares videos of the rehearsals. The backstage view of the performance reveals how the two public ideologies of science are interwoven.[33] One of the presenters chosen among the "ordinary people" is unable to use the steps leading to the stage. The woman asks if she can "enter through the side door, like professor Dulbecco." But after a brief consultation with his team, Fazio rejects this possibility.

Dulbecco is therefore one of us only up to a certain point. His "special" status confers a privilege that cannot be extended to ordinary people without a risk of collision between different roles and dignities. An exception can be made for him but not for ordinary mortals.

On February 25, another festival guest, Teo Teocoli, says to *Repubblica* that "It would not be the same at all [as Dulbecco appearing here] if Teocoli turned up at a scientific congress in Pasadena; I would be incapable of impressing anybody."

The main effect of Dulbecco's presence seems to be to offer a sort of ready-made resource for rhetoric rather than scientific content—a rhetoric that was not necessarily understood but useful in other contexts. The presence of a Nobel laureate on stage at the Ariston Theater of Sanremo creates a sort of "aura" effect: science is back on the agenda (in a rhetorical sense, not for its content), with the protagonists and commentators using themes and metaphors from the scientific world to describe people and situations during the festival or simply to make jokes.

During the opening evening, for example, Fazio introduces the French model Laetitia Casta as a "genetic masterpiece." The comedian Teo Teocoli describes Dulbecco as a "magnet, charged with energy and enthusiasm." During the last evening, an impersonator of Rita Levi-Montalcini calls Fazio "a masterpiece by Dulbecco: much more handsome than Dolly the sheep."

What about the audience? Does the show act as a "propaganda for science," as Dulbecco claims? A survey of a representative sample of the Italian public immediately following the festival shows little impact. Most Italians were aware of Dulbecco's participation and identified him as a scientist, even those who did not follow the festival on television, but few can pinpoint his precise area of expertise: 22 percent believed him to be a physicist or a chemist. And more people who watched the festival every night were unsure of his area of expertise (18 percent) than among nonspectators.[34]

Overall, rather than conveying information and stimulating an interest in science, as Dulbecco hoped, the presence of a Nobel laureate at the Sanremo Festival seems to have had two rather different effects. On one hand is the "prestige transfer." After a few rather unsuccessful seasons, the festival, led by Fabio Fazio, invites a Nobel Prize winner as part of a strategy to enhance the profile of the event. On the other hand, perhaps more surprisingly, the presence of Renato Dulbecco signals and embodies the increasing relevance of science and scientific themes.

SOME NOBELS ARE MORE NOBEL THAN OTHERS?

Does the name of John Bardeen mean anything to you? Or Frederick Sanger? Or Barry Sharpless? Probably not. Well, these are the only three scientists in history to have received the Nobel Prize twice in the same discipline.

Bardeen, an American, was awarded the Nobel Prize for physics in 1956 with William Shockley and Walter Brattain "for their research on semiconductors and their discovery of the transistor effect." During the celebrations, the King of Sweden politely reprimanded Bardeen for not having brought his children—busy with university exams—to Stockholm. "I will bring them when I win the next Nobel," he replied cheekily. Little did he know. In 1972, he won again, this time with Leon Cooper and John Schreiffer for their theory of superconductivity.

The English scientist Frederick Sanger received the prize for chemistry in 1958 "for his work on the structure of proteins, especially that of insulin." He received the prize again in 1980 with Walter Gilbert, this time "for their contributions concerning the determination of base sequences in nucleic acids." (The other half of the prize went to Paul Berg.)

The American chemist Barry Sharpless was awarded in 2001 "for his work on chirally catalyzed oxidation reactions" and again in 2022 for "the development of click chemistry and bioorthogonal chemistry." These are names that are well known to experts but much less known to the wider public, notwithstanding the fact that they twice won the most prestigious scientific prize and benefited by the visibility that the award guarantees. Yet their visibility is not comparable to that of Watson, Crick, or Einstein, not to mention Marie Curie, the only other scientist to have received two Nobel Prizes (in two different science disciplines: physics in 1903 and chemistry in 1911). And speaking of Watson and Crick, why are they remembered, but the third laureate that year, Maurice Wilkins, never mentioned? Wilkins, also a British citizen, was also excluded from the offer of a CBE (Commander of the Most Excellent Order of the British Empire), which was made to Crick after the award (Crick refused the offer). A similar situation occurred to Robert Curl Jr., the third laureate for chemistry in 1996 "for the discovery of fullerenes" with Harold Kroto and Richard Smalley. From the moment the prize recipients were announced, Curl recollects, "I was very rarely mentioned in the press; Harry [Kroto] in Europe and Rick here in England did most of the interviews."[35] And who can remember the person who shared the 1903 Nobel with Marie and Pierre Curie? It was the French physicist Henri Becquerel.

Which laureates are most familiar to the public and why? Some information is provided by an in-depth analysis of Nobel Prize coverage by the *Corriere della Sera*. From a survey of articles dedicated to the prizes awarded between 1901 and 2009, the most visible Nobel laureates (mentioned more than ten times) are Albert Einstein, Rita Levi-Montalcini, Carlo Rubbia, Enrico Fermi, Giulio Natta, Renato Dulbecco, Riccardo Giacconi, and Emilio Segrè.

Of the 566 Nobel Prizes that appear in the *Corriere*, each laureate is mentioned twice on average, meaning that the overwhelming majority of the Nobel laureates do not make a meaningful impression on the media or the public except for a rare quote at the time of the award.

Looking at articles that discuss the Nobel more generally, although in a limited, more recent period, the situation seems to be much the same, except that attention seems even more focused on a few figures. Dulbecco, Levi-Montalcini, and Rubbia are by far the most frequently mentioned. They are followed by Natta, Giacconi, and the first two non-Italians: Einstein and Watson. The first three alone account for over two-thirds of the mentions. "For whoever has, to him more shall be given": the Matthew effect also works with Nobel Prizes.

A first obvious result therefore has to do with the nationalistic element in public discourse about the Nobel Prize (see chapter 3). The media outlets of any given country most often celebrate Nobel laureates and

Table 6.1 The twenty most-mentioned Nobel laureates in articles about the awards and prize giving in the *Corriere della Sera*, 1901–2009

	Articles	
	Number	Percentage
Levi-Montalcini	19	5.8
Rubbia	18	5.5
Fermi	14	4.2
Natta	13	3.9
Dulbecco	11	3.3
Einstein	11	3.3
Giacconi	11	3.3
Segrè	11	3.3
Bovet	9	2.7
Luria	9	2.7
Salam	8	2.4
Cohen	7	2.1
Marconi	7	2.1
Planck	7	2.1
Van der Meer	7	2.1
Bloch	6	1.8
Curie, Marie	6	1.8
Golgi	6	1.8
Watson	6	1.8
Wilson	6	1.8
Total articles	330	100.0
Total mentioned Nobel laureates	566	
Total quotes	1152	
Average number of citations per single Nobel	2.04	

inventions linked to the country of origin. It should be noted that some of the foreign names cited were in fact joint winners of the prize with Italians (Cohen with Levi-Montalcini, Van der Meer with Rubbia) or associated somehow with Italy (Salam worked in Trieste). To confirm this tendency, the three most cited Nobel laureates by the English newspaper *The Times* during the same timeframe were the two British scientists Paul Nurse and Tim Hunt and one American (James Watson) who shared the prize with two British scientists.

Table 6.2 The Nobel laureates most mentioned by the *Corriere della Sera* (in general), 2000–2002

	Number of articles
Dulbecco	93
Levi-Montalcini	85
Rubbia	61
Natta	18
Giacconi	16
Einstein	11
Watson	7
Golgi	6
Horvitz	5
Jacob	5
Brenner	5
Prusiner	5
Segrè	5
Sulston	5
Marconi	3
Average quote per Nobel laureate	4.0

Source: L. Beltrame, "L'immagine del premio Nobel nella stampa quotidiana. Un confronto internazionale" (The Image of the Nobel Prize in the Daily Press: An International Comparison), degree thesis in sociology, University of Trento, 2003.

The archives of the *Corriere della Sera* show the impact of the prize on the visibility of scientists, especially the Italian ones. The attention given to Renato Dulbecco (cited in only four articles before he was awarded the prize) increased more than one hundred-fold (to 542 articles) after he received the prize; Carlo Rubbia was cited in thirty-eight articles before the prize and 877 times after; Rita Levi-Montalcini in fifty-four articles prior to the prize and 1,741 afterward. Frederick Sanger was a complete stranger to the Italian press before his prize (zero articles), and two Nobel prizes were not sufficient to lift him to visibility (he appears in just four articles: he is cited practically only in connection with the announcement and prize ceremony). The situation is quite similar if one looks at the different editions of the *Enciclopedia Italiana Treccani*.[36] The profiles of Dulbecco and Rubbia were published in the *Enciclopedia* for the first time after they won the prize, in the 1978 and 1992 appendixes, respectively, while Rita Levi-Montalcini was already cited before the award. Golgi was quoted in the first edition of 1929–1939 but had only very limited space and no portrait. In the same edition, the Nobel Prize won by Albert Einstein was not even mentioned in his profile; it is possible that he was already so famous that the prize took on a secondary importance as an accessory that was taken for granted (the *Enciclopedia* did, however, mention his "Zionist propaganda" activity).

This tendency to focus on the Nobel Prizes awarded to people from one's own country merits two considerations. First of all, even the "national Nobel winners" are not visible in equal measure; on the contrary, there are obvious differences among them. Henry Nielsen and Keld Nielsen reached the same conclusion after analyzing the space dedicated to the biographies of the first Danish Nobel laureates in the daily press and the national encyclopedias: some are accorded more print space than others. Similarly, some foreign laureates stand out for their celebrity in the Italian press, while others remain largely forgotten.

If this is the situation in the press, what is the level of recognition of Italian Nobel laureates among the public? According to data by the *Osservatorio Scienza Tecnologia e Società* (Science in Society Monitor, the key study on public perception of science in Italy, conducted for more than 20 years), 85 percent of Italians know what the Nobel Prize is. More than half of those (58 percent and 57 percent, respectively) are aware that

Carlo Rubbia and Guglielmo Marconi received a Nobel, while barely more than two out of ten know that Giulio Natta and Emilio Segrè have won the prize. The substantial overlap of notoriety and the Nobel Prize is confirmed by the fact that that over six out of ten interviewees mistakenly thought that a very visible scientist like Margherita Hack had won the Nobel Prize. Finally, when showed some images of famous laureates, almost all of those interviewed (91 percent) were able to recognize Albert Einstein, and two out of three recognized Marie Curie in a portrait of her as a young woman.[37]

There is no doubt that some Nobel laureates are therefore "more Nobel than others." But it is difficult to give an unequivocal answer as to why some are more famous than others or why some have had a stronger impact on our memories or on public perception. It is also necessary to take into account factors beyond the prize that affect scientists' visibility—positively or negatively.

We could schematically divide the Nobel Prizes for sciences into three categories: the scientists who were famous prior to receiving the prize (those who have "given the prize more than what they received from it"), those who became famous after having won the prize, and those who remained more or less unknown to the public after receiving it. Table 6.3 summarizes these three "ideal types," giving a few names as examples.

Among the factors that contribute to the visibility of a Nobel Prize, one can mention the following:

- National identity. It has been seen that, in general, greater visibility is given to recipients from one's own country and that a sense of national competition is fed when the prize is obtained.

Table 6.3 Scientists and visibility linked to the Nobel

Scientists famous prior to the Nobel	Einstein Marconi Marie Curie
Scientists who became famous thanks to the Nobel	Dulbecco Watson Crick
Scientists who remained relatively unknown to the wider public even after having received the Nobel	Sanger Bardeen Curl Wilkins

- The type of discovery for which someone was awarded (for example, Marie Curie's discoveries with radioactivity, the structure of DNA for Watson and Crick). Even its name can have importance. According to some commentators, in the case of Kroto and his colleagues, the choice to name the carbon structure of their discovery "fullerene" in honor of the structures created by the architect Buckminster Fuller turned out to be particularly auspicious.[38] Nevertheless, so-called eponymy (the attribution of a name to natural phenomena or laws) does not guarantee its permanence in collective memory. Camillo Golgi, who gave his name to the "Golgi apparatus," is not particularly well remembered even by Italians.[39]
- The personality, private life, and physical appearance of the laureate, which will be seen in more detail in the next chapter (the most emblematic case from this point of view is certainly that of Einstein).
- The active participation of the awardee in public debate, as well as in occasions and topics outside his scientific expertise (as was the case with Dulbecco starting from the end of the 1990s, with his presidency of the commission bearing his name on stem cells in 2000 and even with his participation in the Sanremo Festival in 1999). This participation in the public debate can take the form of great success as a popularizer. This was the case of Alexis Carrel, a surgeon of French origin who was awarded the Nobel Prize for medicine in 1912 and the author of the international bestseller *Man, the Unknown* (1935). Other Nobel who had great editorial successes were Erwin Schrödinger (*What Is Life?*, 1945), Jacques Monod (*Chance and Necessity*, 1970), and especially James Watson, who wrote the much discussed and successful autobiographical bestseller *The Double Helix* (1968).
- The possibility for the media to make connections between current events of public importance and the figure of the laureate: for example, the so-called brain drain, investments in research, and the role of women in science. This is the case with Rita Levi-Montalcini, especially in recent years and following her appointment as senator for life.

Some of these factors can be combined and therefore mutually reinforce each other. The theme of identity or national competition can overlap with themes of particular public relevance or a specific political situation. This was the case for the German scientists who were awarded the prize at the end of the two wars. It also happened more recently in Italy when the prize awarded to scientists of Italian origin who had been working abroad for many years (such as Riccardo Giacconi, Nobel for physics in 2002) was interpreted within the context of the discussion on "brain drain" and more generally of insufficient Italian investment in research.

More generally, what guarantees the individual laureate a passport to popularity and collective memory is the capacity to become part of the narratives that characterize the prize and to be relevant to these narratives. Two of these, the narrative of the genius and of the national hero, have already been analyzed in previous chapters. The next chapter is dedicated to a third main narrative.

7

THE BODY OF THE (LAUREATE) SCIENTIST

Divine is the illusion. This is a saint. It is like this for all saints, fundamentally unprepared, or rather, hopeless. That is how we pray today. As always. To spend time with the most talented does not mean coming closer to the absolute.
—Carmelo Bene, *Our Lady of the Turks*

THE FACE OF EINSTEIN

On November 3, 1921, the *Corriere della Sera* reported on a conference attended by Einstein in Bologna:

Professor Einstein with his theory of relativity wants to reach only the minds of physicists and mathematicians. But if, as happens to all inventors, he had taken pride in departing from his field of knowledge, he would have discovered the most beautiful example of relativity above his head in this hall of the Archiginnasio. We spectators, squeezed along the wall behind him, were staring at a painting up near the ceiling of a grand and regal Madonna and child; and beneath the Madonna was a bust, of marble perhaps, of King Vittorio Emanuele the Second; and under the bust, finally, Einstein alive and smiling that, as a Jew, he *spoke to us more or less of paradise*, or at least of the place where Dante placed paradise, in the sky, among the stars. [. . .]

It is unlikely that Einstein had ever before spoken under the protection of the Madonna.

A beautiful head, pale and Semitic, of a yellow pallor that, from afar and against the dark backdrop of the school blackboard, appeared more swollen than fat. The hair is shiny, curly, and black hair, with a few stray gray strands. Beneath a black moustache, red and swollen lips; round eyes; the eyebrows short, high, and set apart from one another, so that when he raises his eyes to the sky searching for a word, his whole face takes on a look of surprise, an almost ecstatic expression that moves souls. But what wins them over is his boyish expression. He is a calm, polite, great boy, happy to play with ideas, worlds, and infinity. His fat, shiny, soft hands escaping from his sleeves that are too tight and too short; his limited, shy gestures; his speech, slow, lisping, and hesitant; his ready, sincere, and jovial smile—all enchance his boyish expression. And does he not tell the story of his theory with the grace with which fairytales are told and with the faith of children in those fairytales?—There were seven stars, like this . . . , and on the blackboard he draws seven white stars and then a circle, the sun. The world is finite, but it is unlimited . . . —and he laughs as if in the game he has just thrown the ball so far up into the sky that not even he is able to see it any longer.

This serenity, this boyish freshness, is the magic with which he is able to keep this great public captivated and, they say, in love. It is the reason for which we feel a similarity between this relentless mathematician who bears the name of a stone and a poet. He has the same wings, the same thirst for infinity, the same faith in the reality of dreams—I mean of hypotheses: the same faith in the absolute and also in relativity. To me it seemed that the Christian Madonna—from her throne that is eternal, I know, but painted and therefore fleeting and relative—looked over him benevolently.

This summarizes, via one of its greatest and most celebrated protagonists, the main aspects of one type of public iconography of scientists. His physical aspect, starting from his face, is transfigured. The "beauty of his head" is extended to the rest of his figure and is amalgamated with his moral qualities (education, serenity, sincerity, and boyish innocence), taking on a quality that is not only ethical but also esthetic (he is defined as "a poet") and even "ecstatic." His face "moves souls" and is compared to the figure of a king and even religious symbols. This is the image of the "scientist as saint" that Einstein symbolized during the years of his popularity with the great international public and his winning the Nobel Prize.[1]

Understanding how the Nobel placed itself in the stream of this iconography is one of the keys to understanding how the prize has contributed to shaping the public image of scientists over the past century.

THE SECULAR WORSHIP OF SCIENTISTS

The public iconography of scientists, particularly during the nineteenth century, coincided with the social and institutional success of science and with the rising idea in many societies that religious saints should be joined in adulation by lay figures of strong symbolic importance, who were increasingly expected to be celebrated with biographies, portraits, and monuments. This iconography appropriated stylistic elements from religious iconography, applied them to the figures of scientists, and turned them into references and symbols of the new secular sensitivity and the increasingly significant social and cultural role of science.

Frederic Harrison, an English student of the French positivist philosopher Auguste Comte, put together the *New Calendar of Great Men* in 1892 in which every month was associated with the name of a representative figure. The physiologist Xavier Bichat was the "saint" of the last month, dedicated to modern science. Each month was divided into four weeks dedicated to four scientific subjects (astronomy, physics, chemistry, and biology). As in religious calendars in which each day is assigned to a saint, Harrison's calendar gave a scientist's name and brief biography to each date. A few years earlier, the French chemist Gaston Tissandier, founder of the journal *Nature*, had offered a list of sumptuously illustrated "martyrs of science" (1879) to the public.

These celebrations tended to underline the intellectual and disembodied, almost ascetic character of scientific personalities and stressed their detachment from materiality. Recall the final images of Darwin, where all that is left for the public to see are "his beard, his hat, and his eyes" or the recurring images of his now empty study, but "full of signs of his mind. [. . .] Darwin, as a physical presence, had almost disappeared [. . .] his intellect appeared to the public in an almost completely disembodied form."[2]

This iconography is similar to that used for figures recognized as "repositories of truth and values, whether religious, scientific, philosophical, or artistic" within "conceptions of virtuous and sacred knowledge attached to special people inhabiting special bodies"[3] but is also clearly influenced by religious iconography that praised disinterest in dietary pleasures and even the "wonderful abstinences" attributed to some of its most famous representatives. Numerous stories and anecdotes emphasize Newton's indifference about food and how he "would leave his food waiting for two hours on the table," "often ate his cold dinner for breakfast," and "fattened his cat with all the leftovers from his tray." No less picturesque, especially in popular biographies, were the stories of Pasteur, who continues his microscope observations, regardless of the reminders of the wife who calls him to lunch.[4]

Thus, in representations of scientists—by underlining character traits such as sobriety, frugality, and even indifference to material needs and pleasures—modern culture "ingeniously reworked"[5] the Christian precepts of moderation and frugality. A well-known anthologist of the lives of saints, Adrien Baillet, was chosen to be the biographer of Descartes by some of his disciples. Thanks to biographies such as John Conduitt's, the figure of Newton "was constructed like that of a scientific saint in line with 'catholic guidelines.'" Vincenzo Viviani's biography of Galileo described with admiration how little he slept, his indifference for life at court, and his rejection of all other potential distractions from work. These works delineated figures of "ascetic intellectuals who redeemed with their own upright conduct the heterodox scope of their thinking."[6]

Although he does not explicitly refer to scientists in his description of "men of letters" as heroes, Carlyle draws a comparison with "the same function which the old generations named a man prophet, priest, divinity for doing."[7]

These themes were particularly apparent in writings about Michael Faraday, who was, during his lifetime, "revered like a scientific saint."[8] After heated discussions, his statue—proposed for Westminster Abbey or St. Paul's Cathedral—was finally placed "on a plot that had already been sanctified by the scientific community" in the atrium of the Royal Institution, where it still stands to this day.[9] In 1931, on the centenary of the discovery of electromagnetic induction, the statue was placed at the center of an exhibition that had many parallels to religious celebrations and attracted many visitors, further publicizing the discovery.[10] Alessandro Volta (1878) and Luigi Galvani (1879) also had monuments dedicated to them in Italy.

Some of these commemorations set off debates and even open conflict between science and religion. An emblematic event in the Trentino region involved a bust of the naturalist Giovanni Canestrini (1835–1900), to whom we owe the first Italian edition of Darwin's *Origin of Species*, on which he collaborated with Leonardo Salimbeni. The decision to honor him with a plaque in 1901 and then a commemorative bust occasioned a violent dispute that highlighted the way in which the scientist's figure and public relevance were defined, especially by his strongest supporters.

His biography, published at the beginning of October, was similar in style to a hagiography: the poor childhood, the carpenter father, the financial support for his studies from an uncle who was secretary to the archbishop of Gorizia, his precocious vocation for the study of nature. The story of his last lesson on embryology at the University of Padua was memorable: "the professor, who was listened to religiously, spoke with difficulty in a hoarse voice albeit with his usual clarity and his simple, noble style."

"The great Canestrini deserves a statue not a bust," wrote the famous medical doctor and criminologist Cesare Lombroso in the *Alto Adige* of September 26–27, 1901. The following year, at the inauguration of a bust dedicated to Canestrini by students from the Trentino region, the anthropologist Lamberto Moschen (one of his collaborators from the University of Padova) made a famous speech. In it, he underlined that "Canestrini was one of the few Italians who adhered immediately to the new doctrine [of Darwinism] and became a fervent apostle of the theory of evolution."[11] This idea was emphasized from the moment the debate was reignited in the summer of 1902, when the bust of Canestrini was erected by the Society of Students from Trentino. The marble bust representing the scientist was inaugurated on September 14 with a solemn ceremony.[12] The occasion was inevitably favorable for reassessing Canestrini's biography not only in the scientific context but also for its moral, civil, and even political import:

> His life was sanctified by civil and domestic virtues and entirely dedicated to work, often parched by harsh battles, because in openly proclaiming his opinions and convictions he came up against ancient prejudices and deep-rooted mistakes. [. . .] This man, says Moschen, had a gentle soul and the most generous heart you could possibly imagine [. . .]. In his heart, science and family only had one competitor, his homeland that he loved deeply. [. . .] During the fights that await you, oh young ones, always remember this valorous man who fought for his faith, this tireless worker and champion for freedom and progress; and may the highly civil role that is accomplished today thanks to you be the prelude for fortunate battles for freedom, independence, and our country's prosperity.[13]

THE BODY OF THE SCIENTIST AND ITS RELICS

On October 5, 1895, a few weeks before Alfred Nobel drafted his will, a grand collective rite was celebrated in Paris. A silent procession walked behind a hearse to the cathedral of Notre Dame. In addition to the French president and minister for education, important figures from all over Europe (from Grand Duke Constantine of Russia to Prince Nicola of Greece) came to pay their respects at the state funeral of the great scientist Louis Pasteur, famous for his advancements in the fight against infectious diseases. "Pasteur Is Eternal," ran the headline in one paper. After his death, Pasteur was depicted with muses at his feet or as a "saviour with a halo above his head, sometimes bedecked with wings, suffering the little children to come unto him."[14] His embalmed body was buried in a mausoleum at the Institut Pasteur, which some have called a "sanctuary" dedicated to the scientist and where some of his personal effects and other objects (his desk, his instruments) are displayed. Until recently, institute personnel met in the mausoleum twice a year on the anniversary of Pasteur's birth and death. Nobel laureate François Jacob recalls:

> the descent into the crypt began, in Indian file, in hierarchical order: the director and the board; council; then the department heads [. . .], then the technicians and assistants, finally, the cleaning women and lab boys. Each went slowly down some steps before passing in front of the tomb [. . .]. At the entrance, over the whole of the vault, mosaics depicted, in the manner of scenes from the life of Christ, those from the life of Pasteur: sheep grazing, chicken pecking, garlands of hops, mulberry trees, grapevines, representing the treatment of anthrax, chicken cholera, the diseases of beer, of the vine, of the silkworm. And at the summit, the supreme image, the struggle of a child with a furious dog, to glorify the most decisive battle, that against rabies. In the center, on the cupola's pendentives, four angels with outspread wings three representing the theological virtues of Faith, Charity, and Hope, the fourth, judged fitting by turn-of-the century scientism, representing Science.[15]

This "worshipping" of Pasteur was reinforced by a perception of his frail physique, especially during the last days of his life.

Scientists have a "double body," like the sovereigns studied by E. H. Kantorowicz (1957): a natural body that is physical and mortal and another that is consecrated and immortal, created by their accomplishments and the public's imagination.[16] The "immortal body" is magnified as if to compensate for the frail physical body. The myth of Pasteur was consolidated during the last years of his life, when his "brutal and despotic" character (according to his detractors) was substituted by a fragile figure who provoked compassion, as one visitor recalled:

His leg and left arm, smitten by apoplexy, are somewhat stiff, and [. . .] he drags one foot much like a wounded veteran. Age, illness, the heavy labours of so many years, the bitterness of conflict, the intense passion for his work, and, lastly that prostration which follows triumph, have combined together to make a grand thing of his face. Weary, traversed with deep furrows, the skin and beard both white, his hair still thick, and nearly always covered with a black [skull] cap, the broad forehead wrinkled, seamed with the scars of genius, the mouth slightly drawn by paralysis, but full of kindness, all the more expressive of pity for the sufferings of others [. . .] and above all, the living thought which still flashes from the eye beneath the deep shadow of the brow—this is Pasteur as he appeared to me: a conqueror, who will someday become a legend, whose glory is as incalculable as the good he has accomplished.[17]

The importance taken on by scientific figures and their physical aspect is confirmed by a long series of "scientific relics" that have been kept, exhibited, and in some cases even secretly stolen. The middle finger of Galileo's right hand is exhibited today the Galileo Museum in Florence next to the following verses (originally in Latin) by Tommaso Perelli:

This is the finger, belonging to the illustrious hand
That ran through skies
Pointing at the immense spaces and singling out new stars,
Offering to the senses a marvelous apparatus
Of crafted glass.

Many other relics of Galileo have been kept, including his fifth lumbar vertebra, which was also used for posthumous diagnoses of the pathologies that affected the scientist.[18]

The heads and particularly the brains of scientists have naturally been the focus of the greatest attention. The head of the anatomist Antonio Scarpa is at the Museum of the University of Pavia and even touted as a macabre "tourist attraction" on Tripadvisor. In Turin, the skeleton and brain of Carlo Giacomini, the director of the Cabinet of Anatomy, are preserved at the Anatomy Museum of the University; he died in 1898 and was very active in the conservation of brains. Giacomini was the one to ask that his bones be kept at the Institute of Anatomy, "where I have spent the best years of my youth and to which I have dedicated all my efforts. I would like my brain to be preserved with my method and kept in the museum with all the others."[19]

But the most famous case is certainly that of Einstein's brain. In 1955, Thomas Harvey, a doctor at Princeton Hospital, removed Einstein's brain with no authorization during his autopsy. His family and the executors of his will found out after this body had been cremated and his ashes scattered according to his own last wishes. They left his brain to doctor Harvey on the condition that he share it with other scholars purely for scientific research. Later Harvey, who was fired from the hospital due to his actions, sectioned Einstein's brain in approximately two hundred small "slices" and returned some to Einstein's doctor. He sent others to researchers who wished to study one of the most famous brains in history to try to identify some material clues to the extraordinary intellect of its owner. But attempting to extrapolate from the dimension of the inferior parietal lobe or the density of his neurons often led to controversial conclusions.

American journalist Michael Paterniti chronicled the adventures of the leftover parts of the brain in his book *Driving Mr. Albert: A Trip across America with Einstein's Brain*. In 1997, Paterniti accompanied the now elderly Harvey on a trip to hand over the remainder of Einstein's brain to his nephew, who refused it. Today, approximately three hundred photos of Einstein's brain matter are available in digital format thanks to a special smartphone application.[20]

Posthumous devotion to certain scientific figures has also been facilitated by the conservation and "veneration" of some objects linked to them: work instruments, private objects, portraits. On the centenary of the Nobel Prize, the newly established Nobel Prize Museum exhibited a few objects that had belonged to different laureates: personal notebooks and notes, surgical instruments, the microscope belonging to Santiago Ramon y Cajal, containers with Hafnium salts belonging to George de Hevesy, and test tubes used by Francis Peyton Rous.[21]

"PRINCES WHO PAY HOMAGE TO MOLECULES": THE NOBEL AS A CONSECRATION RITUAL

The Nobel Prize, its ceremonies, and its rituals fit into—and contribute to strengthening and consolidating—this narrative.

On one hand, the prize focuses attention on individual figures who acquire strong collective symbolism. These are geniuses whose recognition metonymically incarnates public and collective admiration of their scientific discovery. Through the prize, science—an entity that is abstract and with which the wider public is not very familiar—acquires a face and a body. On the other hand, what makes the Nobel Prize special compared to other prestigious (and more financially rewarding) prizes is the elaborate ceremony that characterizes it. If the check were sent anonymously by post, as suggested in 1907 by the Swedish writer Verner von Heidenstam (Nobel laureate for literature in 1916), the prize would lose one of its key elements.

The "Royal touch"—the presentation of the prize by the King of Sweden that officially transforms a brilliant scientist into a Nobel laureate—has been compared to the "consecration" and coronation ceremonies that turn sovereigns into the "magic-working kings" studied by Marc Bloch, to whom medieval beliefs attributed the capacity to heal scrofula with touch.[22] Numerous analogies have also been made between the Nobel ceremony and religious canonization.

One scholar of canonization ceremonies wrote, "To the outside world, canonization is rather like the Nobel Prize: no one really knows why one candidate is chosen over another or who—apart from the Pope—does the selecting."[23] In both realms, there is an elaborate investigation to acquire information and prepare documentation; there are "promoters" of canonization just as scientists send in nominations for the Nobel; there are figures whose task is to look into the merit of each candidacy; at least one specific "miracle" has to be performed by the candidate in question (or in scientific words, a specific discovery or invention that merits the prize); questions such as "is the candidate's reputation for [. . .] extraordinary virtue founded on fact?" must be answered.[24]

There are many examples in which the authority and influence of Nobel Prizes are compared to that of the highest religious figures or representatives, starting with Enrico Fermi's famous appellation as the "Pope of physics" (and Emilio Segrè's as the "Abbot") by their colleagues in via Panisperna. A sort of "papal dogma of infallibility" was attributed to the Nobel laureate for chemistry (1965) Robert Woodward; they said of him that "until the Pope had given his approval to something, it didn't exist."[25] The physicist Maria Goeppert, Nobel recipient in 1963 for having discovered the presence of a series of shell structures in the nucleus similar to the layers of an onion, was nicknamed "the Onion Madonna" by Wolfgang Pauli. "For a while now I have been treated like an oracle," Carlo Rubbia jokingly complained to the *Corriere della Sera* (November 16, 2000). The new London-based mega-laboratory for biomedicine directed by the Nobel for medicine Paul Nurse was called "Sir Paul's Cathedral" by *Nature*.

And then, as always, there is Einstein, "the new messiah, the First Knowledgian and Supreme Head of the Vast Physical Universe."[26] When Einstein moved to the United States, his friend and colleague Paul Langevin defined it as "an event of extraordinary importance, as would be the transfer of the Vatican from Rome to the New World."[27]

Websites and maternity services nowadays offer future parents names inspired by Nobel Prize laureates: "If your baby is bound for greatness, try starting with a name inspired by one of the world's best thinkers."[28]

The award ceremony of the Nobel Prize can be interpreted as a rite during which "a certain sacredness is bestowed upon the individual, that is expressed and confirmed by symbolic acts."[29] The central element of these ceremonies is *deference*,

by which appreciation is regularly conveyed [. . .] to a recipient or to something of which this recipient is taken as a symbol, extension or agent.[30]

Thus, with the ceremony of the Nobel and the elaborate rituals during the week of the awards ceremony, deference is expressed to the chosen scientists and, through them, to science. The driver and diplomatic personnel available to the laureate for the whole week, the banquets and toasts in their honor, the proximity to members of royal families, the "royal touch" when handed the prize, the medal with its reference to nobility and military glory: they are all part of this unusual ritual.

The recollections and memories of laureates often underline this aspect. "Each laureate was accompanied by a princess from the royal family," says the *Corriere della Sera* of the 1933 ceremony. The chemist John Polanyi (Nobel Prize in 1986) expressed this ironically during his speech at the banquet:

Your Majesties, Your Royal Highnesses, ladies and gentlemen. [. . .] I know of no other place where princes assemble to pay their respects to molecules. [. . .] Because of you, our wives hesitate for just an instant before summoning us to wash the dishes.[31]

The ritual element is amplified by the media attention on the event, which is broadcast live by the Swedish state television channel SVT, with videos and images transmitted across the world. The Nobel Prize ceremony is the sort of media event that scholars call "a coronation." Like royal weddings, royal coronations, or the funerals of great personalities, the ceremony of the Nobel is symbolically centered on the values of "tradition and continuity." By widely publicizing the event, we collectively celebrate and reaffirm the tradition of science and the continuity of its cultural and social role.[32]

Deference is also expressed to the body of the scientist, particularly if it is a frail one. When Giulio Natta received the Nobel Prize for chemistry in 1963, he was already suffering from Parkinson's disease. He reached his seat with great difficulty and the help of his son. When the master of ceremonies invited him to go down the stairs toward the king, he almost fell over. When he understood the situation, King Gustav of Sweden, who was eighty-one at the time, made an exception to the rigid protocol and "went up the stairs two at a time and standing there on the platform, handed over the medal to the invalid laureate. The audience burst into prolonged applause."[33]

On the other hand, the focus on the frailty of the body is characterized by a certain iconography of the scientist whose physical weakness and unassuming appearance reflects his total dedication to knowledge and a pure spiritual vocation. Frederick Banting, recipient of the Nobel Prize for the discovery of insulin in 1923, was well known for his clumsiness and his tendency to stumble over his words. These characteristics contributed to making cases like his a "perfect story," combining the figure of a scientist as a genius, hero, and saint:

> the wounded veteran, the failing small-city doctor, the great idea at night, nothing but discouragement from the establishment [...], grinding poverty, imaginative experiments under the worst conditions [...] and then brilliant spectacular success.[34]

The other face of deference is *demeanor*, meaning

> that element of the individual's ceremonial behavior typically conveyed through deportment, dress, which serves to express [...] that he is a person of certain [...] qualities.[35]

Through demeanor, the person receiving deference confirms that he is worthy of it. The award ceremony for the prize is one of the main occasions during which composure can be expressed. The laureates wear a dinner jacket and, during the rehearsal for the ceremony, are briefed in detail on their movements: how to receive the diploma and shake the king's hand, how to take a bow (three times: to the king, to the laureates from the previous years, and to the public).

Gaffes and incidents still occur: John Bardeen, on the occasion of his first prize, wore an embarrassing shirt and a waistcoat bearing a green stain due to a washing mishap. Eric Cornell, who received the prize for physics in 2001, fumbled repeatedly before shaking the sovereign's hand and completely forgot to take a bow. In most of these cases, the loss of composure is justified or tempered by the "special" nature of the scientists, whose distraction or clumsiness is justified and compensated for by their exceptional intellect.

But Nobel Prize laureates are not expected to show composure only during the ceremony. One of the main characteristics that demonstrates the laureates' composure is their emphasis on modesty. Modesty is a sort of counterbalance to the competition for originality and recognition, competition that the prize expresses at the highest level and that, as we have seen, can lead to harsh disputes between scientists. Modesty is expressed, for example, by recognizing the contributions made by collaborators and predecessors. This has been a widespread attitude since the beginnings of modern science, made famous by Newton's famous words: "If I have seen further, it is by standing on the shoulders of giants."[36] Thus, during the prize giving, official speeches, and interviews to the press, many recipients emphasize the role of their collaborators or mentors:

"There is one thing that I would like to say," declared Emilio Segrè after the announcement of the Nobel Prize for physics: "My only regret is that Fermi is not alive. It is on days like this that we most feel the absence of those dear people who have been linked to our work. And it is to Enrico Fermi that I owe early guidance in my studies and in nuclear physics research."[37]

Peter Agre, recipient of the Nobel Prize for chemistry in 2003, emphasized this aspect in an ironic and paradoxical way: "I didn't do this work, the young people in the lab did it. I just made the coffee and sharpened the pencils." The Nobel laureate for medicine Charles Richet (1913) liked to quote these words of a colleague: "I possess every good quality, but the one that distinguishes me above all is modesty."[38]

The ritual of alternating speeches by representatives of the Nobel committees, who praise the individual contribution of each laureate, with those by the laureates, who in turn redistribute the merit, enables the Nobel event to celebrate both personal successes and scientific activity as a whole.[39]

Modesty, humility, simple attire and habits that contrast with the exceptional nature of the discoveries: these are the recurring themes that run through the public iconography of the Nobel Prize. The portrait of the Curie couple is memorable—simple dress, a modest suburban home, and even "simple bodies":

> The Curries [sic] live very isolated, outside the Latin Quarter near the ramparts, in an area that, for Parisians of the boulevards, is farther away than Siberia. The coachmen don't even know the name of the street. Mr. Currie is in his forties, tall, slim, with very simple manners and the modesty of a true scientist. He presents himself as a good, gentle, and very shy man. His wife came to Paris to study when she was young. She distinguished herself in the science faculty where she obtained a diploma and where she met Mr. Currie, who asked her to marry him. They had met in a physics and chemistry laboratory, and from the union of these two simple bodies was born a baby girl called Irene.[40]

Because "once a Nobel always a Nobel" and because this new status invests scientists not just with recognition but also with moral authority, the laureates' behavior must be in line with these expectations even after they have won the prize.

The most serious offenses—or the ones that are perceived as such—to this "Nobel composure" have been called (again recalling the consecration process) "sins against the prize." There have been requests, for example, to change of the name of streets named after Nobel laureates such as Alexis Carrel, suspected of being a Nazi sympathizer after a series of documents emerged during the 1990s. Requests have even been made to revoke the prize for frontal lobotomy from Egas Moniz. There has been outrage over the conviction for sexual assaults on minors of the Nobel laureate for medicine Carleton Gajdusek (in his obituary, the British newspaper *The Guardian* mentioned "the rare distinction of being a Nobel prizewinner and a convicted child molester"). In 1988, on the occasion of the fiftieth anniversary of Richard Kuhn's Nobel (refused at the time due to the Nazi boycott of the prize), the Austrian postal service had planned to issue a stamp in honor of the scientist. Following the discovery of embarrassing documents in which Kuhn stated his support for annexation to

Germany, the proposal was set aside, though revived again years later for the twenty-fifth anniversary of his death.

As significant as consecration from the prize is, it can be ruinous to go against the demeanor that it requires of Nobel laureates. Following his declarations in 2007 in which he criticized the intelligence of African Americans, the Nobel laureate James Watson lost many consultancy contracts and opportunities to hold lectures. "I have become a *nonperson*," he declared: "no one wants to admit I exist."[41]

One cannot be a "bit Nobel," just as one cannot be a "bit saintly." The alternative to the extreme visibility that the prize guarantees—if one denies it by not behaving—is "invisibility." Watson's term to describe his invisibility and absence of social consideration, *nonperson*, is the same used by the sociologist Goffman to describe those who are present during social interactions but treated as if they are not (a typical example is that of the servant in some households at some times in history).[42]

If the role of Nobel laureate consecrated by the ceremony is contradicted, one cannot go back to being "a simple scientist"; nor can one completely disappear. That is how Watson justified his decision to sell his Nobel medal at auction, thus ridding himself symbolically of his status.[43]

"SCIENCE AS A BRIDE": THE EXCEPTIONALITY AND MORAL EQUIVALENCE OF SCIENTISTS

In 1942, while introducing his definition of ethos in science, the sociologist Robert K. Merton noted that to the scientist has been attributed

> A passion for knowledge, idle curiosity, altruistic concern with the benefit to humanity, and a host of other special motives [. . .]. By implication, scientists are recruited from the ranks of those who exhibit an unusual degree of moral integrity. There is, in fact, no satisfactory evidence that such is the case; a more plausible explanation may be found in certain distinctive characteristics of science itself.[44]

For Merton, therefore, it is not the moral characteristics of scientists as individuals but rather the hinges supporting science as an institution that guarantee its functioning and development. It is the "the public and testable character of science" that "has contributed to the integrity of the men of science," not vice versa.[45] It is not the scientists who are special but science as an institution, which is why historically it has often been at odds with other institutions (political, economic, and religious) that have other objectives.

This argument, referred to as "the moral equivalence of scientists," was not a new concept when Merton mentioned it, but it certainly could not be taken for granted.

The public image of scientists since the birth of modern science had, on the contrary, been forged from an idea of the scientist having distinctive moral motivations and qualities. Science was born as a "vocation" well before it became a profession, striving to "the Glory of the Great Author of Nature, and to the Comfort of Humanity"—in the famous words from the will of Robert Boyle, one of the founders of the Royal Society, inspired by the great theorist and apologist of modern science, Francis Bacon. For Bacon, knowledge should not be

a courtesan, for pleasure and vanity only, or as a bondwoman, to acquire and gain to her master's use; but as a spouse, for generation, fruit, and comfort.[46]

During the Renaissance, the term *virtuous* (from the Latin *virtus*, "capacity," "value") defined the moral impulse of "rational artists" and the first natural philosophers as

a program for relating man to the world as perceiver and knower and agent in the context of his integral moral, social, and cosmological existence [...] a common style in the mastery of self, or nature and of mankind alike by the rational anticipation of effects.[47]

This style was perfectly incarnated by Galileo and other key figures of the scientific revolution.

The "exceptional morality" of scientists, their conformity with their vocation, and their indifference toward material interests and mundane disputes seems obvious to the philosopher Adam Smith in his *Theory of Moral Sentiments* (1759):

Mathematicians and natural philosophers, [...] are almost always men of the most amiable simplicity of manners, who live in good harmony with one another, are the friends of one another's reputation, enter into no intrigue in order to secure the public applause.[48]

This exceptional nature is often traced back by scientists themselves to their field of study: it is the investigation into nature as a divine creation that elevates and guarantees scholars' moral quality. In 1916, the physicist and editor in chief of the scientific journal *Nature*, Sir Richard Gregory, still stated that

> the conviction that devotion to the study of nature gives courage and power to those who possess it; it is the divine afflatus which inspires and enables the highest work in science.[49]

The complete dedication required by the search for truth and understanding of nature is often described as acting to the detriment of life and private relationships of scientists. When Guglielmo Marconi split from his fiancée Josephine Holman, the newspaper *Il Resto del Carlino* (February 4, 1902) gave an explanation that almost literally echoed Bacon's notion of "science as a bride":

> the end of the idyll—which attracted general attention in the United States—was due to this: the fiancée, and particularly her mother, wanted the marriage to be celebrated quickly. Marconi, on the other hand, was firm in his idea that marriage should wait until the proof of wireless transatlantic transmissions have reached such a degree of perfection that he could afford the time and leisure a not so short holiday.

Nevertheless, from the beginning of the twentieth century, the perception of a moral equivalence of scientists started to change. Science started to be perceived as a profession, not a vocation. Scientific research became more organized and institutionalized; rapid technological developments made huge impressions on the public; and later some scientists' involvement with military initiatives damaged scientists' reputation for "exceptional morality."

After the Second World War and especially following the so-called Sputnik effect—the immense repercussions of the Soviet space success that led American elites to boost support for research and recruitment of young people in scientific disciplines during the Cold War—scientists' supposedly exceptional morality started to be perceived as an obstacle. "The Einstein complex," as the physicist Luis Alvarez called it, risked relegating science to an activity for eccentrics and a few people with a clear vocation. Communication started to focus on the normality and the ordinary nature of scientists and scientific enterprises: "Hero—and Human Being" was the headline of a special edition of *Life* magazine in 1964.[50]

And it is precisely into these changes that the Nobel Prize inserted itself.

Are scientists special, or are they ordinary human being beings? The prize turned out to be the perfect narrative context for managing and balancing these two registers. By elevating individuals (never more than three, even when celebrating results obtained by larger groups of researchers), by rewarding them and putting them in the spotlight, the Nobel committee confirms that they are special. Otherwise, why reward them in particular?

But at the same time, they are placed within a ceremonial and communications context that departs from traditional ones in the world of research and science. The prize's attendant sums of money inevitably lead to attention on practical aspects and material benefits. By exposing honorees to questions, requests, and curiosities of all sorts, a more "human" dimension emerges that connects them to daily life. This continuous alternating between two registers is one of the keys to understanding the enduring success of the prize and its role in defining the public image of science and scientists over more than a century.

The alternating and combination of registers often appears in the words of Nobel laureates themselves. David Baltimore, Nobel for medicine in 1975, describes a fairly typical experience: "People keep writing to me via email to ask, 'What is the meaning of life?' And they want me to respond quickly via email!" The Nobel laureate is a recognized moral authority, but the wider public expects to have an intimate relationship with him that is "lay" and informal and typical of scientific communication.

The ambiguous, precarious balance between the exceptional and the ordinary moral nature of scientists is explained more dramatically by Max Born (physics 1954): "I now regard my former belief in the superiority of science over other forms of human thought and behaviour as a self-deception."[51]

We can consider these two aspects in terms of the classic sociological distinction between "front-stage" and "back-stage."[52] On the front-stage, meaning in public or in the ceremonial contexts linked to prize giving, we keep to a repertoire of moral exceptionality by emphasizing qualities such as disinterest in and indifference to material benefits. In private, in informal conversations with colleagues, aspects that contradict this repertoire can emerge. James Watson's autobiography, *The Double Helix* (1968), caused quite a stir in this respect. Opposed by his colleague Crick, accepted and then rejected by Harvard University Press because it was considered embarrassing, the biography did not hesitate to underline the harsh competition between rival scientists and their burning ambition to win the Nobel. The main interpretation of Watson's book by the public and the press was, in fact, that "scientists are human, after all [. . .] scientist are all too human [. . .], they can be boastful, jealous, garrulous, violent [and even] stupid."[53]

A more sophisticated way to reconcile the tension between the different registers is the one chosen by the Nobel for medicine Jacques Monod in his bestseller *Chance and Necessity*: "Modesty befits the scientist, but not the ideas that inhabit him and that he *is under the duty* of upholding" (emphasis in the original).[54] Moral exceptionality is thus not typical of scientists as such but rather of their relationship with scientific concepts and activity.

Even the physical dimension is implicated in this alternating of registers. On one hand, the body of the Nobel laureate is special, as required by tradition. Einstein's face, particularly the portrait of him sticking his tongue out, has become emblematic of the unconventional and playful genius. His brain, described as "a modern day myth" by Roland Barthes, and his reduction to pure intelligence meant that "just by being his colleague, every physicist wore a halo of humility." There are even stories of male and female admirers fainting in surprise when coming face to face with him. By dodging the ritual of the prize-giving ceremony, the symbolic figure of Einstein avoided all mundane contamination. There are incredible stories of his Japanese tour, which coincided with the awarding of the prize. The German ambassador in Tokyo reported:

His trip through Japan resembled a triumphal procession. Whereas the visit of the Prince of Wales and Field Marshall Joffre were attended by royal and military pomp, advanced detailed planning and officious play-up in the press, there was none of this during Einstein's reception; instead the participation of the entire Japanese people was in evidence, from the highest dignitary to the rikshaw coolie; spontaneous, without preparation and window dressing! When Einstein arrived in Tokyo, there were such crowds at the station that the police could do nothing to control this life-and-limb-threatening crush of people. [. . .] All eyes were riveted on Einstein; everyone wanted at least to shake the hand of the most famous man of present times. An admiral in full uniform forced his way through the rows of people, when up to Einstein and said "I admire you," and thereupon, departed.[55]

A Japanese newspaper wrote:

The great and noble light in his eyes was an incredible wonder to me. His gentle attitude of total absorption in his research almost moved me to tears.[56]

The moral nobility of the scientist reflects off the body and even its adornments, as was the case with Rita Levi-Montalcini. At the awards ceremony of 1986, the scientist was described as being "gentle and noble" with her "white head of hair [. . .] in her long, classic, almost nineteenth-century style evening dress" while she walked confidently toward King Carl Gustav."[57]

Even the scientific discovery takes on a human body during the prize-giving evening:

On Christmas Eve 1986, the NGF [nerve growth factor, the discovery for which Montalcini was awarded the prize] appeared [. . .] in public under large floodlights, in the splendor of a room festively decorated, in the presence of the royals of Sweden, the princes, and dames in magnificent evening gowns and gentlemen in tuxedos. Wrapped in a black mantle, the NGF bowed before the king and, for a moment, lowered the veil covering her face. We recognized each other in a matter of seconds when I saw her looking for me among the applauding crowd. She then replaced her veil and disappeared as suddenly as she had appeared.[58]

The thaumaturgic virtues and the exceptional nature of Nobel laureates' bodies come back to life in the much debated "Nobel Prize sperm bank," a project that was launched in 1980 in California and promised potential mothers "the most selective genes." Worried perhaps by the furious controversies that exploded soon after the announcement (with accusations of racism and of a potential eugenic risk), only three Nobel laureates actually contributed their sperm, and only one (William Shockley, awarded the prize for physics in 1956) publicly admitted to having donated. The project was shut down at the end of the 1990s after claiming that it had provided sperm for the birth of over two hundred children, of which none were to receive the Nobel Prize.[59]

The body of the Nobel laureate is also a heroic body that, in the name of science, never gives in to its frailty. Of the Nobel for medicine in 1914, Robert Bárány, the press underlined that although he was a prisoner of the Russians and suffering from malaria, he was working on a new scientific project. No less emblematic was the example of the prize winners for medicine who put their lives at risk by experimenting with new therapies on themselves or their family members. Recall that Banting injected insulin into himself and his assistant. Barry Marshall, Nobel Prize in 2005, demonstrated the role of the *Helicobacter pylori* in ulcerous pathologies by experimenting with it on himself after taking a sample from a patient. "I had succeeded in being infected. I had proven my thesis," he declared in his biography during the prize giving. Some "relics" of his martyrdom, including the vial from which he drank the infected solution, have been exhibited at the Nobel Prize Museum in Stockholm. Tu Youyou (medicine, 2015) tested the antimalarial efficiency of a substance contained in the *Artemisia annua* plant on herself: "as head of the research group it was my responsibility." Gerhard Domagk, one of the scientists who could not receive the award because of the German boycott, experimented with sulphonamide in antibacterial therapy on his daughter, who actually recovered from a bad wound infection with a needle.

The traditional iconography of temperance, moderation, sobriety, and even misogyny is renewed through Nobel laureates. According to George Beadle, Nobel Prize for medicine in 1958:

You too can win Nobel Prizes. Study diligently. Respect DNA. Don't smoke. Don't drink. Avoid women and politics. That's my formula.

The *Corriere della Sera* on October 17, 1986, emphasized the contrast between the prestige of the prize and the frugality of Levi-Montalcini's lifestyle, calling her award a "Nobel worth 300,000 lira per month":

— Professor, is it true that you have a research contract at the CNR that pays you only 400,000 lira per month?
— No, look, for that matter, it's a little less, roughly 300,000 per month. But I am seventy years old, and that's the way it should be. It is right that young researchers are paid better. As for me, I live with very little: my diet is based on boiled rice, frozen fish, and vegetables. I am absolutely fine, and this is plenty for me.

The same goes for modesty and humility. One day, John Bardeen was asked, "What exactly do you do?" by his golf partner of many years. Apparently, Bardeen had never mentioned that he had won two Nobel Prizes for physics. When Lawrence Bragg moved from Cambridge to London, he wanted to continue his gardening hobby, so he put an ad in the paper for a job as a part-time gardener. Only after a few months did one visitor ask the house owner "what Sir Lawrence was doing pruning the roses."⁶⁰ "A humble genius" was how the Nobel laureate for medicine Frederick Banting was described.

Some laureates even show indifference to the prize, such as Samuel Ting, Nobel Prize for physics in 1976, interviewed by the *Corriere della Sera* on December 15:

How does it feel to be catapulted into the forefront of science at the age of forty? "No feeling, I didn't feel anything in particular. My daughter is the one who collects the articles and interviews concerning me, I am too busy with work."

An enormous number of anecdotes address the moment when scientists receive news of the prize: anecdotes that often emphasize stupor, modesty, simple manners, sobriety, and the ordinariness of daily life. Peter Higgs says he found out from a neighbor on the street and received the official message only later. John Bardeen says he was so surprised that he dropped the eggs he was cooking for his family. Giulio Natta claims to have found out while in Alfred Nobel's beloved Sanremo, where he had gone to rest.

Scientists' total dedication to their work emerged with the story of the announcement of the prize for physics in 2010 to Konstantin Novoselov. He was busy with an experiment when he received the call: "So basically you are telling me I should interrupt my experiments now?" he asked. "I'm afraid so," replied the Foundation spokesperson.⁶¹

Rita Levi-Montalcini's recollections of receiving the Nobel were in the same vein, as she told the *Corriere della Sera* (October 14 and 15, 1986):

It was eleven o'clock in the morning, I was quietly sitting reading the last pages of an Agatha Christie thriller when the phone rang. The Swedish Academy of Sciences was informing me that I had won the Nobel Prize for medicine with my friend and colleague Stanley Cohen. [. . .] I toasted the Nobel with soup: a simple meal of broth and rice followed by a long sleep.

Barry Marshall (medicine, 2005) was contacted by the Nobel Foundation while he was in a bar one day:

Well, we're not [. . .] We're being very careful—we're just having one glass of beer at the moment. And I don't want to appear on television, intoxicated. Dr. Warren and I, we're very moderate in our activities and, usually, one beer is enough to keep us cheerful.⁶²

On the other hand, the ritual and ceremony surrounding the prize highlights the laureates' appreciation for good food, fine wines, elegant clothes, and refined manners, their enjoyment of the company of high society and the glitterati. Photos emerge of scientists dancing and drinking wine during the banquet (one reporter ironically described Röntgen while he "X-rayed his sea bass à la Normande") and the description of Francis Crick in 1962 by the *Corriere della Sera* reads as follows:

[With] his receding hairline due to threatening baldness, with a large nose and a smiling face, he dresses like a perfect Englishman, always in a dark suit, with a bowler hat and the inevitable bright carnation in his left buttonhole. He is very worldly. He talks pleasantly to beautiful women.

This risk of being distracted and even overwhelmed, as scientists, by the pleasantness of the celebrations in Stockholm is recalled by the 1976 Nobel Prize for economics, Milton Friedman: "It's really very, very, very nice for a week. It would corrupt you utterly if it lasted much longer."[63]

The media attention on the families who accompany them—in particular on the husbands and wives, who at times steal the show from the laureates themselves—also contributes to this worldly dimension for the laureates. From the moment they are assaulted by photographers at their arrival with their spouses at the banquet, the laureates discover that "they are no longer alone in the limelight—instead it is their family members who receive all the attention,"[64] meaning that the rituals of deference and the weight of demeanor are extended to spouses and families in equal measure.

At times, a spouse becomes the intermediary through which the Nobel Prize approaches common sense and daily life: an indirect and even *softer* way of balancing the rhetoric of exceptionality with that of normality. "Dear Claudio, Daddy has been awarded the Nobel Prize" is how the wife of Emilio Segrè, Elfriede Spiro, announced the news to their son, according to the *Corriere della Sera*. The son then recalled a hike in the mountains during which his father explained his research to him

with the same simple language used to explain difficult concepts to the uninitiated. [...] In his words science was no longer an abstruse subject; it became poetry, philosophy, the only things that I understand well.[65]

The relationship with laureates' family members and with spouses, in particular, is often an opportunity to emphasize scientists' total dedication to research and consequently the complete delegation of all other prerogatives to their partners. Regarding the use of the prize money, the

Nobel Prize for medicine Arvid Carlsson (2000) said to the press: "I first have to talk to my wife. She is the one who makes these decisions."[66]

There is an anecdote that circulates within the Academy of Sciences about a Nobel Prize laureate who stopped at a gas station with his wife. The wife recognized the station employee and got out of the car to give him a kiss and a hug. When she got back into the car, the woman explained that he was a former boyfriend of hers. The scientist was irritated: "It's embarrassing for me to see you kissing a gas station employee. After all, I am a Nobel Prize laureate!" To which the wife replied: "Yes, you are right, you are a Nobel laureate, but if I had married him, he would have been a Nobel laureate today."

THE FLIP SIDE OF THE NOBEL PRIZE: THE BOXES THAT NO ONE AT THE ACADEMY OF SCIENCES WANTS TO OPEN

To the Royal Swedish Academy of Sciences: I am a sociable person and music lover. I ask your secretary to contact me because I am interested in receiving the Nobel Prize. My phone number is [. . .] (postcard from the United States, 2013)

There are huge boxes in the basement of the Academy of Sciences, full of correspondence of this nature: letters, manuscripts, and illustrated texts in which authors lay claim to discoveries, results, and merits that should be considered worthy of the most important scientific prize. Nothing gets thrown out, yet nothing is made public unless it is discussed or taken into consideration, since the institutions who assign the prizes evaluate only nominations sent in by those who are entitled to do so within the expected timeframe and required procedures. This is how the reputation of the prize is guaranteed: by archiving material that, if it were made public, could fuel needless curiosity and disputes. At the same time, this endless flow of requests is proof of the popularity of the prize, famous and desired even by those who know only its name and not much else.

One implicit way of exorcising this material and therefore protecting the dignity of the prize and its recipients is the existence of initiatives such as the Ig Nobel Prize.

Created in 1991 by the journal *Annals of Improbable Research*, the prizes are assigned to "results that first make people laugh and then make them think." The structure is based on that of the Nobel Prizes, copying the fields that are awarded the prize (physics, chemistry, medicine, peace, literature, economics) and including some others (mathematics, biology, but also reproduction, management, perception, etc.). The prizes go to real research projects, published in respected journals and carried out by researchers from institutions that are often prestigious, but the projects are characterized by hilarious goals or conclusions. For example, the prize for medicine in 2016 went to the article "Itch Relief by Mirror Scratching: A Psychophysical Study," whereby if one has an itch on the left-hand side of the body, one can feel relief by looking at oneself in the mirror and scratching one's right-hand side (and vice versa). For physics, one prize went to a study on why dragonflies prefer to settle on black tombstones in cemeteries. The memory of Alfred Nobel, deformed and veiled at times, is often invoked during the prize ceremony, which takes place in September at Harvard University and at which "real" Nobel laureates are invited to present the prizes.

Only one researcher to date has won both the Nobel and the Ig Nobel: the scientist Andre Geim, of Russian origin. He received the Ig Nobel in 2000 "for using magnets to levitate a frog." In 2010, he went on to be awarded the Nobel Prize for physics for his research on graphene. The Ig Nobel Prizes are a means for science to reach the public by revealing its funny, ridiculous side without questioning its dignity or that of its main protagonists, made famous by the Nobel Prize. It is a sort of exhaust valve that enables science to maintain the two registers that characterize the contemporary image of the scientist on two different yet compatible levels. Geim declared to the BBC, "Frankly, I value my Ig Nobel as much as my Nobel Prize; for me, the Ig Nobel is proof that I can take a joke. A bit of self-deprecation always helps."[67] The eternal modesty of the Nobel Prize laureate, with a touch of irony.

Irony can also be a way to maintain a certain "distance from the role" of being a Nobel laureate and the composure that this requires, which can result in being perceived as excessively rigid.[68] Carlo Rubbia explained his "recipe for the Nobel" to his colleagues at Harvard: "You people think it's hard to win a Nobel Prize, but it's easy. Trivial. Just put protons and antiprotons in a box, and shake them up, and then collect your prize."

More generally, irony is a tool widely used to being together the two narratives—that of moral exceptionality and that of scientists as "human, too human." The Nobel laureate for medicine François Jacob commented sarcastically on the infamous Nobel Prize sperm bank: "Some have praised the sperm of the Nobel Prize winners. Only someone who does not know Nobel laureates would want to reproduce them like that."[69]

In the inevitable stories of the moment in which the news was received, scientists often refer to the possibility of jokes on the part of friends and colleagues. "But I don't have friends with such a perfect Swedish accent, so I started to believe it was true," said the Nobel for chemistry in 2012 Brian Kobilka. In his typical irreverent style, Kary Mullis answered the phone, "I'll take it! I just wanted to make sure there were no doubts about it."[70]

Martin Chalfie, Nobel Prize for chemistry in 2008, missed the call from Stockholm because he was still asleep in New York due to the time difference:

Ah. This is a sort of ridiculous situation. But sort of funny. I woke up at ten after six, and I realized that they must have given the Prize in Chemistry, so I simply said, "Okay, who's the schnook that got the Prize this time?" And so I opened up my laptop, and I got to the Nobel Prize site and I found out I was the schnook![71]

When James Watson received the Nobel Prize for medicine in 1962, one of the telegrams congratulating him was signed "Gly like Glycerine," a nickname given to the future Nobel laureate for physics Richard Feynman in the exclusive RNA Tie Club founded in 1954 by Watson and the physicist George Gamow "to solve the enigma of the structure of RNA."

Perhaps alluding to the fact that Watson was still single or to the fact that his great ambition had finally been rewarded or to the Baconian metaphor of "science as a bride," the telegram read: "And there he met the beautiful princess, and they lived happily ever after."

EPILOGUE: GENIUSES, HEROES, AND SAINTS—HOW THE NOBEL PRIZE (RE)INVENTED THE PUBLIC IMAGE OF SCIENCE

[Jules] Verne got up from his chair. "À qui ai-je l'honneur?," he asked.
"I am Science," his visitor replied.
And from that day on, that austere lady became for him what Troy had been for Homer: a source of inspiration.
—A. Savinio, *Narrate, uomini, la vostra storia*

In 1906, the American economist and sociologist Thorstein Veblen noted that "on any large question which is to be disposed of for good and all the final appeal is by common consent taken to the scientist" and defined science's place in modern civilization: "Quasi lignum vitae in paradiso Dei, et quasi lucerna fulgoris in domo Domini" (Like a tree of life in the garden of God and like a lamp in the house of the Lord). Veblen then wondered: "How has this cult of science arisen?"[1]

The answer to this question is partly provided by the Nobel Prize. Naturally, it is difficult to say to what degree the prize has helped turn science and scientists into a key institution of contemporary society and culture and to what extent it has reflected, enhanced, and in some ways even driven the development and consolidation of the social role of science and its protagonists.

What I hope I have shown here are the power and importance of the three narratives that characterize the prize and its public dimension: the scientist as genius, the scientist as national hero, and the scientist as saint.

The narrative of the genius emphasizes the creativity of the scientist, the intellectual exceptionality, which is reflected in a solitary and romantic ideal that is very effective from the point of view of public image.

The narrative of the national hero enables the Nobel-winning scientist to speak as the voice of a nation, replacing and sublimating national tensions and rivalries through a more peaceful and noble competition.

The narrative of the saint *embodies* (literally, as we have seen) the scientist's moral exceptionality, revising the traditional idea of the man of science as a secular ascetic.

The three narratives complement, balance, and reinforce each other. "In a sacred idiom, scientific discovery is divine inspiration; in a secular idiom it is spontaneous and serendipitous."[2] When combined, they define the Nobel Prize as "the right prize at the right moment."

In an age when research was already becoming a more complex, organized, and unavoidably depersonalized activity, the narrative of the genius allowed individual contributions and faces to be put into focus.

At a time in history when national rivalries were finding more peaceful expression through events such as the Olympic Games and international exhibitions such as the world's fair and when science was beginning to be seen as the expression of a nation's power, the prize was an extraordinary opportunity to express "political rivalry through other means," favored in part because of its location in neutral Sweden. When the "moral exceptionality" attributed to scientists began to be questioned and research started to be socially defined as a job more than a vocation, the prize gave a new language to scientific virtues like modesty, humility, and total dedication—*body and soul*—to the scientific enterprise.

In a research context characterized by the increasingly rapid proliferation, reinterpretation, and inevitable obsolescence of results, the Nobel Prize allows a few key figures (and the developments and achievements that they represent) to be fixed in the collective memory and rendered "immortal," eternalizing them in a sort of *secular pantheon*. The balancing of complementary narrative registers—creative and moral exceptionality versus proximity to the everyday and worldly dimension—helps define behavioral models that are distant and yet accessible and comprehensible, at least from a human point of view: themes such as emancipation from conditions of exclusion or the injustice that characterizes many of the stories of those who won the Nobel and those who did not, particularly between the two world wars.

The elaborate ritual and ceremony and the "royal touch" that consecrates the laureate and "princes and princesses paying tribute to the molecule" symbolically condense the deference that society and politics bestow on science.

Together, the three narratives also convey a certain interpretative flexibility to their protagonists. Within the broad boundaries of the demeanor expected of Nobel Prize laureates, there are "various ways of being a Nobel," emphasizing one narrative over another or balancing the narratives on the basis of different historical contexts or specific personal trajectories.

From this point of view, the Nobel laureate as a genius, the Nobel laureate as a national hero, and the Nobel laureate as a saint are simply ideal types, continuously recombined and reelaborated by individual scientists. The adaptability and versatility of the narratives allow each Nobel laureate to perform a range of interpretations and various degrees of "distance from the role" and allow for the growing articulation of the scientist's role in contemporary society and the different "social circles" of reference: other academic disciplines, business and technological applications, the media, and the general public. In a capacity of focused yet ecumenical narrative, the Nobel laureate can be, in turn (or even simultaneously), a poetic and disembodied thinker, a technological leader, an entrepreneur, a military consultant, a popularizer.[3]

This combination of registers recalls that of the prize's founder, Alfred Nobel: chemist, inventor, entrepreneur, first defined the "merchant of death" and then a benefactor of humanity. Even in its more controversial choices (or perhaps because of them), the Nobel Prize has been able to represent the intrinsic ambivalence of science, whose power and practical implications began to appear as evident as its cognitive ambitions, starting with the activities of Alfred Nobel himself.

One could wonder whether this increasing articulation and plasticity have not become a symptom of the prize's weakness. The fragmentation of what was once called the "scientific community"; the increasingly diverse interpretation of the role of researcher; the growing porousness and permeability of the world of research with regard to other areas, actors, and models of activity such as business, media communication, the mobilization of citizens and patients: they seem to now question some of the organizational, social, and cultural premises on which the prize was originally based. Compared to the famous group photograph of the Solvay Conference in 1927, with its compact and glorious crowd of Nobel Prize–winning scientists, the contemporary image seems more unfocused, changing, or simply impossible to condense into a single image.

The role of the prize and its recipients can also be analyzed in the context of "the decline of public intellectuals." According to some scholars, one of the underlying elements of this decline is the growing tendency of experts and academics to comment on topics and matters outside their personal realm of competence—a characteristic dynamic of the public visibility of scientists, as we have seen, that is common to at least a certain interpretation of the Nobel Prize's role.[4]

Thus, perhaps the Nobel Prize has become a victim of its own success. By fostering the personalization and celebrity of its protagonists, the prize might have helped undermine the original foundations of its own identity and reputation, such as its competence. By enhancing visibility and transforming it into all-round celebrity, it openly negates scientific virtues such as humility and modesty, which are the essential institutional counterbalance to the celebration of the individual. The Matthew effect, if pushed to its extreme, can turn against its own beneficiaries and against science as an institution.

Nonetheless, these contemporary considerations and challenges do not make the role of the Nobel Prize any less significant in shaping an image of science and of the scientist, which, from a social and cultural point of view, has been—and still is—a reference point.

For the general public, science largely remains an abstract and inscrutable entity. The Nobel Prize has helped give it a face and a body and provided a repertoire of stories that still deserve to be told. Together with the story that summarizes them and makes them all possible, the only intuition that is celebrated every year: the greatest invention by a man who owned 355 patents.

In the words of a Nobel laureate, spoken during the ceremonial banquet:

We applaud you, therefore, for your discovery, which has made a memorable contribution to civilization—I refer, Your Majesties and our Swedish hosts, to the institution of this unique prize, for which we, in the company of many others, thank you.[5]

APPENDIX: ALL THE NOBEL LAUREATES IN THE SCIENCES, 1901 TO 2024

	Physics	Chemistry	Physiology or Medicine
1901	Wilhelm Conrad Röntgen	Jacobus Henricus van 't Hoff	Emil Adolf von Behring
1902	Hendrik Antoon Lorentz and Pieter Zeeman	Hermann Emil Fischer	Ronald Ross
1903	Antoine Henri Becquerel • Pierre Curie and Marie Curie, née Sklodowska	Svante August Arrhenius	Niels Ryberg Finsen
1904	Lord Rayleigh (John William Strutt)	Sir William Ramsay	Ivan Petrovich Pavlov
1905	Philipp Eduard Anton von Lenard	Johann Friedrich Wilhelm Adolf von Baeyer	Robert Koch
1906	Joseph John Thomson	Henri Moissan	Camillo Golgi and Santiago Ramón y Cajal
1907	Albert Abraham Michelson	Eduard Buchner	Charles Louis Alphonse Laveran
1908	Gabriel Lippmann	Ernest Rutherford	Ilya Ilyich Mechnikov and Paul Ehrlich
1909	Guglielmo Marconi and Karl Ferdinand Braun	Wilhelm Ostwald	Emil Theodor Kocher

(continued)

	Physics	Chemistry	Physiology or Medicine
1910	Johannes Diderik van der Waals	Otto Wallach	Albrecht Kossel
1911	Wilhelm Wien	Marie Curie, née Sklodowska	Allvar Gullstrand
1912	Nils Gustaf Dalén	Victor Grignard • Paul Sebatier	Alexis Carrel
1913	Heike Kamerlingh Onnes	Alfred Werner	Charles Robert Richet
1914	Max von Laue	Theodore William Richards	Robert Bárány
1915	Sir William Henry Bragg and William Lawrence Bragg	Richard Martin Willstätter	—
1916	—	—	—
1917	Charles Glover Barkla	—	—
1918	Max Karl Ernst Ludwig Planck	Fritz Haber	—
1919	Johannes Stark	—	Jules Bordet
1920	Charles Edouard Guillaume	Walther Hermann Nernst	Schack August Steenberg Krogh
1921	Albert Einstein	Frederick Soddy	—
1922	Niels Henrik David Bohr	Francis William Aston	Archibald Vivian Hill • Otto Fritz Meyerhof
1923	Robert Andrews Millikan	Fritz Pregl	Frederick Grant Banting and John James Rickard Macleod
1924	Karl Manne Georg Siegbahn	—	Willem Einthoven
1925	James Franck and Gustav Ludwig Hertz	Richard Adolf Zsigmondy	—
1926	Jean Baptiste Perrin	The (Theodor) Svedberg	Johannes Andreas Grib Fibiger
1927	Arthur Holly Compton • Charles Thompson Rees Wilson	Heinrich Otto Wieland	Julius Wagner-Jauregg

APPENDIX

	Physics	Chemistry	Physiology or Medicine
1928	Owen Willans Richardson	Adolf Otto Reinhold Windaus	Charles Jules Henri Nicolle
1929	Prince Louis-Victor Pierre Raymond de Broglie	Arthur Harden and Hans Karl August Simon von Euler-Chelpin	Christiaan Eijkman • Sir Frederick Gowland Hopkins
1930	Sir Chandrasekhara Venkata Raman	Hans Fischer	Karl Landsteiner
1931	—	Carl Bosch and Friedrich Bergius	Otto Heinrich Warburg
1932	Werner Karl Heisenberg	Irving Langmuir	Sir Charles Scott Sherrington and Edgar Douglas Adrian
1933	Erwin Schrödinger and Paul Adrien Maurice Dirac	—	Thomas Hunt Morgan
1934	—	Harold Clayton Urey	Georg Hoyt Whipple, George Richards Minot, and William Parry Murphy
1935	James Chadwick	Frédéric Joliot and Irène Joliot-Curie	Hans Spemann
1936	Victor Franz Hess • Carl David Anderson	Petrus (Peter) Josephus Wilhelmus Debye	Sir Henry Hallett Dale and Otto Loewi
1937	Clinton Joseph Davisson and George Paget Thomson	Walter Norman Haworth • Paul Karrer	Albert von Szent-Györgyi Nagyrápolt
1938	Enrico Fermi	Richard Kuhn	Corneille Jean François Heymans
1939	Ernest Orlando Lawrence	Adolf Friedrich Johann Butenandt • Leopold Ruzicka	Gerhard Domagk
1940	—	—	—
1941	—	—	—
1942	—	—	—

(continued)

	Physics	Chemistry	Physiology or Medicine
1943	Otto Stern	George de Hevesy	Henrik Carl Peter Dam • Edward Adelbert Doisy
1944	Isidor Isaac Rabi	Otto Hahn	Joseph Erlanger and Herbert Spencer Gasser
1945	Wolfgang Pauli	Artturi Ilmari Virtanen	Sir Alexander Fleming, Ernst Boris Chain, and Sir Howard Walter Florey
1946	Percy Williams Bridgman	James Batcheller Sumner • John Howard Northrop and Wendell Meredith Stanley	Hermann Joseph Muller
1947	Sir Edward Victor Appleton	Sir Thomas Robinson	Carl Ferdinand Cori and Gerty Theresa Cori, née Radnitz • Bernardo Alberto Houssay
1948	Patrick Maynard Stuart Blackett	Arne Wilhelm Kaurin Tiselius	Paul Hermann Müller
1949	Hideki Yukawa	William Francis Giauque	Walter Rudolf Hess • António Caetano de Abreu Freire Egas Moniz
1950	Cecil Frank Powell	Otto Paul Hermann Diels and Kurt Alder	Edward Calvin Kendall, Tadeus Reichstein, and Philip Showalter Hench
1951	Sir John Douglas Cockcroft and Ernst Thomas Sinton Walton	Edwin Mattison McMillan and Glenn Theodore Seaborg	Max Theller
1952	Felix Bloch and Edward Mills Purcell	Archer John Porter Martin and Richard Laurence Millington Synge	Selman Abraham Waksman
1953	Frits Zernike	Hermann Staudinger	Hans Adolf Krebs • Fritz Albert Lipmann
1954	Max Born • Walther Bothe	Linus Carl Pauling	John Franklin Enders, Thomas Huckle Weller, and Frederick Chapman Robbins

APPENDIX

	Physics	Chemistry	Physiology or Medicine
1955	Willis Eugene Lamb • Polykarp Kusch	Vincent du Vigneaud	Axel Hugo Theodor Theorell
1956	William Bradford Shockley, John Bardeen, and Walter Houser Brattain	Sir Cyril Norman Hinshelwood and Nikolay Nikolaevich Semenov	André Frédéric Cournand, Werner Forssmann, and Dickinson W. Richards
1957	Chen Ning Yang and Tsung-Dao (T. D.) Lee	Lord (Alexander R.) Todd	Daniel Bovet
1958	Pavel Alekseyevich Cherenkov, Ilya Mikhailovich Frank, and Igor Yevgenyevich Tamm	Frederick Sanger	George Wells Beadle and Edward Lawrie Tatum • Joshua Lederberg
1959	Emilio Gino Segrè and Owen Chamberlain	Jaroslav Heyrovský	Severo Ochoa and Arthur Kornberg
1960	Donald Arthur Glaser	Willard Frank Libby	Sir Frank Macfarlane Burnet and Peter Brian Medawar
1961	Robert Hofstadter • Rudolf Ludwig Mössbauer	Melvin Calvin	Georg von Békésy
1962	Lev Davidovich Landau	Max Ferdinand Perutz and John Cowdery Kendrew	Francis Harry Compton Crick, James Dewey Watson, and Maurice Hugh Frederick Wilkins
1963	Eugene Paul Wigner • Maria Goeppert Mayer and J. Hans D. Jensen	Karl Ziegler and Giulio Natta	Sir John Carew Eccles, Alan Lloyd Hodgkin, and Andrew Fielding Huxley
1964	Charles Hard Townes, Nicolay Gennadiyevich Basov, and Aleksandr Mikhailovich Prokhorov	Dorothy Crowfoot Hodgkin	Konrad Bloch and Feodor Lynen
1965	Sin-Itiro Tomonaga, Julian Schwinger, and Richard P. Feynman	Robert Burns Woodward	François Jacob, André Lwoff, and Jacques Monod
1966	Alfred Kastler	Robert S. Mulliken	Peyton Rous • Charles Brenton Huggins

(continued)

	Physics	Chemistry	Physiology or Medicine
1967	Hans Albrecht Bethe	Manfred Eigen, Ronald George Wreyford Norrish, and George Porter	Ragnar Granit, Haldan Keffer Hartline, and George Wald
1968	Luis Walter Alvarez	Lars Onsager	Robert W. Holley, Har Gobind Khorana, and Marshall W. Nirenberg
1969	Murray Gell-Mann	Derek H. R. Barton and Odd Hassel	Max Delbrück, Alfred D. Hershey, and Salvador E. Luria
1970	Hannes Olof Gösta Alfvén • Louis Eugène Félix Néel	Luis F. Leloir	Sir Bernard Katz, Ulf von Euler, and Julius Axelrod
1971	Dennis Gabor	Gerhard Herzberg	Earl W. Sutherland Jr.
1972	John Bardeen, Leon Neil Cooper, and John Robert Schrieffer	Christian B. Anfinsen • Stanford Moore and William H. Stein	Gerald M. Edelman and Rodney R. Porter
1973	Leo Esaki and Ivar Giaever • Brian David Josephson	Ernst Otto Fischer and Geoffrey Wilkinson	Karl von Frisch, Konrad Lorenz, and Nikolaas Tinbergen
1974	Sir Martin Ryle and Antony Hewish	Paul J. Flory	Albert Claude, Christian de Duve, and George E. Palade
1975	Aage Niels Bohr, Ben Roy Mottelson, and Leo James Rainwater	John Warcup Cornforth • Vladimir Prelog	David Baltimore, Renato Dulbecco, and Howard Martin Temin
1976	Burton Richter and Samuel Chao Chung Ting	William N. Lipscomb	Baruch S. Blumberg and D. Carleton Gajdusek
1977	Philip Warren Anderson, Sir Nevill Francis Mott, and John Hasbrouck van Vleck	Ilya Prigogine	Roger Guillemin and Andrew V. Schally • Rosalyn Yalow
1978	Pyotr Leonidovich Kapitsa • Arno Allan Penzias and Robert Woodrow Wilson	Peter D. Mitchell	Werner Arber, Daniel Nathans, and Hamilton O. Smith
1979	Sheldon Lee Glashow, Abdus Salam, and Steven Weinberg	Herbert C. Brown and Georg Wittig	Allan M. Cormack and Godfrey N. Hounsfield

APPENDIX

	Physics	Chemistry	Physiology or Medicine
1980	James Watson Cronin and Val Logsdon Fitch	Paul Berg • Walter Gilbert and Frederick Sanger	Baruj Benacerraf, Jean Dausset, and George D. Snell
1981	Nicolaas Bloembergen and Arthur Leonard Schawlow • Kai M. Siegbahn	Kenichi Fukui and Roald Hoffmann	Roger W. Sperry • David H. Hubel and Torsten N. Wiesel
1982	Kenneth G. Wilson	Aaron Klug	Sune K. Bergström, Bengt I. Samuelsson, and John R. Vane
1983	Subrahmanyan Chandrasekhar • William Alfred Fowler	Henry Taube	Barbara McClintock
1984	Carlo Rubbia and Simon van der Meer	Robert Bruce Merrifield	Niels K. Jerne, Georges J. F. Köhler, and Cesar Milstein
1985	Klaus von Klitzing	Herbert A. Hauptman and Jerome Karle	Michael S. Brown and Joseph L. Goldstein
1986	Ernst Ruska • Gerd Binnig and Heinrich Rohrer	Dudley R. Herschbach, Yuan T. Lee, and John C. Polanyi	Stanley Cohen and Rita Levi-Montalcini
1987	J. George Bedonorz and K. Alexander Müller	Donald J. Cram, Jean-Marie Lehn, and Charles J. Pedersen	Susumu Tonegawa
1988	Leon M. Lederman, Melvin Schwartz, and Jack Steinberger	Johann Deisenhofer, Robert Hubert, and Hartmut Michel	Sir James W. Black, Gertrude B. Elion, and George H. Hitchings
1989	Norman F. Ramsey • Hans G. Dehmelt and Wolfgang Paul	Sidney Altman and Thomas R. Cech	J. Michael Bishop and Harold E. Varmus
1990	Jerome I. Friedman, Henry W. Kendall, and Richard E. Taylor	Elias James Corey	Joseph E. Murray and E. Donnall Thomas
1991	Pierre-Gilles de Gennes	Richard R. Ernst	Erwin Neher and Bert Sakmann

(continued)

	Physics	Chemistry	Physiology or Medicine
1992	Georges Charpak	Rudolph A. Marcus	Edmond H. Fischer and Edwin G. Krebs
1993	Russell A. Hulse and Joseph H. Taylor Jr.	Kary B. Mullis • Michael Smith	Richard J. Roberts and Phillip A. Sharp
1994	Bertram N. Brockhouse • Clifford G. Shull	George A. Olah	Alfred G. Gilman and Martin Rodbell
1995	Martin L. Perl • Frederick Reines	Paul J. Crutzen, Mario J. Molina, and F. Sherwood Rowland	Edward B. Lewis, Christiane Nüsslein-Volhard, and Eric F. Wieschaus
1996	David M. Lee, Douglas D. Osheroff, and Robert C. Richardson	Robert F. Curl Jr., Sir Harold W. Kroto, and Richard E. Smalley	Peter C. Doherty and Rolf M. Zinkernagel
1997	Steven Chu, Claude Cohen-Tannoudji, and William D. Phillips	Paul D. Boyer and John E. Walker • Jens C. Skou	Stanley B. Prusiner
1998	Robert B. Laughlin, Horst L. Störmer, and Daniel C. Tsui	Walter Kohn • John A. Pople	Robert F. Furchgott, Louis J. Ignarro, and Ferid Murad
1999	Gerardus 't Hooft and Martinus J. G. Veltman	Ahmed H. Zewail	Günter Blobel
2000	Zhores I. Alferov and Herbert Kroemer • Jack S. Kilby	Alan J. Heeger, Alan G. MacDiarmid, and Hideki Shirakawa	Arvid Carlsson, Paul Greengard, and Eric R. Kandel
2001	Eric A. Cornell, Wolfgang Ketterle, and Carl E. Wieman	William S. Knowles and Ryoji Noyori • K. Barry Sharpless	Leland H. Hartwell, Tim Hunt, and Sir Paul M. Nurse
2002	Raymond Davis Jr. and Masatoshi Koshiba • Riccardo Giacconi	John B. Fenn and Koichi Tanaka • Kurt Wüthrich	Sydney Brenner, H. Robert Horvitz, and John E. Sulston
2003	Alexei A. Abrikosov, Vitaly L. Ginzburg, and Anthony J. Leggett	Peter Agre • Roderick MacKinnon	Paul C. Lauterbur and Sir Peter Mansfield

APPENDIX

	Physics	Chemistry	Physiology or Medicine
2004	David J. Gross, H. David Politzer, and Frank Wilczek	Aaron Ciechanover, Avram Hershko, and Irwin Rose	Richard Axel and Linda B. Buck
2005	Roy J. Glauber • John L. Hall and Theodor W. Hänsch	Yves Chauvin, Robert H. Grubbs, and Richard R. Schrock	Barry J. Marshall and J. Robin Warren
2006	John C. Mather and George F. Smoot	Roger D. Kronberg	Andrew Z. Fire and Craig C. Mello
2007	Albert Fert and Peter Grünberg	Gerhard Ertl	Mario R. Capecchi, Sir Martin J. Evans, and Oliver Smithies
2008	Yoichiro Nambu • Makoto Kobayashi and Toshihide Maskawa	Osamu Shimomura, Martin Chalfie, and Roger Y. Tsien	Harald zur Hausen • Françoise Barré-Sinoussi and Luc Montagnier
2009	Charles Kuen Kao • Willard S. Boyle and George E. Smith	Venkatraman Ramakrishnan, Thomas A. Stietz, and Ada E. Yonath	Elizabeth H. Blackburn, Carol W. Greider, and Jack W. Szostak
2010	Andre Geim and Konstantin Novoselov	Richard F. Heck, Ei-ichi Negishi, and Akira Suzuki	Robert G. Edwards
2011	Saul Perlmutter, Brian P. Schmidt, and Adam G. Riess	Dan Shechtman	Bruce A. Beutler and Jules A. Hoffmann • Ralph M. Steinman
2012	Serge Haroche and David J. Wineland	Robert J. Lefkowitz and Brian K. Kobilka	Sir John B. Gurdon and Shinya Yamanaka
2013	François Englert and Peter W. Higgs	Martin Karplus, Michael Levitt, and Arieh Warshel	James E. Rothman, Randy W. Schekman, and Thomas C. Südhof
2014	Isamu Akasaki, Hiroshi Amano, and Shuji Nakamura	Eric Betzig, Stefan W. Hell, and William E. Moerner	John O'Keefe, May-Britt Moser, and Edvard I. Moser
2015	Takaaki Kajita and Arthur B. McDonald	Thomas Lindahl, Paul Modrich, and Aziz Sancar	William C. Campbell and Satoshi Ōmura • Tu Youyou

(continued)

	Physics	Chemistry	Physiology or Medicine
2016	David J. Thouless, F. Duncan M. Haldane, and J. Michael Kosterlitz	Jean-Pierre Sauvage, Sir J. Fraser Stoddart, and Bernard L. Feringa	Yoshinori Ohsumi
2017	Reiner Weiss, Barry C. Barish, and Kip S. Thorne	Jacques Dubochet, Joachim Frank, and Richard Henderson	Jeffrey C. Hall, Michael Rosbash, and Michael W. Young
2018	Arthur Ashkin • Gérard Mourou and Donna Strickland	Frances H. Arnold • George P. Smith and Sir Gregory P. Winter	James P. Allison and Tasuku Honjo
2019	James Peebles • Michel Mayer and Didier Queloz	John B. Goodenough, M. Stanley Whittingham, and Akira Yoshino	William G. Kaelin Jr., Sir Peter J. Ratcliffe, and Gregg L. Semenza
2020	Roger Penrose • Reinhard Genzel and Andrea Ghez	Emmanuelle Charpentier and Jennifer A. Doudna	Harvey J. Alter, Michael Houghton, and Charles M. Rice
2021	Syukuro Manabe and Klaus Hasselmann • Giorgio Parisi	Benjamin List and David W. C. MacMillan	David Julius and Ardem Patapoutian
2022	Alain Aspect, John F. Clauser, and Anton Zeilinger	Carolyn R. Bertozzi, Morten Meldal, and K. Barry Sharpless	Svante Pääbo
2023	Pierre Agostini, Ferenc Krausz, and Anne L'Huillier	Moungi G. Bawendi, Louis E. Brus, and Aleksey Yekimov	Katalin Karikó and Drew Weissman
2024	[Winner TK Oct. 7–14, 2024]	[Winner TK]	[Winner TK]

Source: The Nobel Prize, https://www.nobelprize.org.

NOTES

CHAPTER 1

1. H. Schück et al., *Nobel: The Man and His Prizes*, 2nd ed. (Amsterdam: Elsevier, 1962), 18.
2. Schück et al., *Nobel*, 33.
3. Schück et al., *Nobel*, 22.
4. Schück et al., *Nobel*, 23.
5. The text is available at https://www.nobelpeaceprize.org/nobel-peace-prize/history/alfred-nobel-s-will.
6. R. Sohlman, "Alfred Nobel and the Nobel Foundation," in Schück et al., *Nobel*, 54.
7. Sohlman, "Alfred Nobel and the Nobel Foundation."
8. From now on, I shall refer in abbreviated form to the institution that selects the Nobel laureates in chemistry and physics—that is, the Royal Swedish Academy of Sciences (in Swedish: Kungliga Vetenskapsakademien or KVA)—as the Academy of Sciences. This institution should not be confused with the one that selects the Nobel Prizes in literature, which is the Swedish Academy (Svenska Akademien).
9. R. M. Friedman, *The Politics of Excellence: Behind the Nobel Prize in Science* (New York: Times Books, 2001), 16.
10. Sohlman, "Alfred Nobel and the Nobel Foundation," 99ff. Among the many texts dedicated to the biography of Alfred Nobel and the origins of the prize, other than Sohlman's and the other essays in the volume by Schück, one should also refer to K. Fant, *Alfred Nobel: A Biography* (New York: Arcade, 1993) (Swedish original edition 1991); T. Frangsmyr, *Alfred Nobel* (Stockholm: Svenska Institutet, 1996); B. Feldman,

The Nobel Prize: A History of Genius, Controversy and Prestige (New York: Arcade, 2000); Friedman, *The Politics of Excellence*; and the various essays published at the website The Nobel Prize, https://www.nobelprize.org.

11. From "Statues of the Nobel Foundation," at https://www.nobelprize.org/organization/statutes-of-the-nobel-foundation.

12. *Corriere della Sera*, December 12, 1901.

13. *Corriere della Sera*, December 13, 1901.

14. The letter can be found at the official site of the Nobel Prize at "Proclamation Sent to Leo Tolstoy after the 1901 Year's Presentation of Nobel Prizes," The Nobel Prize, https://www.nobelprize.org.

15. E. Crawford, *The Beginnings of the Nobel Institution: The Science Prizes, 1901–1915* (Cambridge: Cambridge University Press, 1984); J. F. English, *The Economy of Prestige: Prizes, Awards and the Circulation of Cultural Value* (Cambridge, MA: Harvard University Press, 2009).

16. Crawford, *The Beginnings of the Nobel Institution*, 192.

17. Until now, only the Peace Prize has been awarded to organizations instead of individuals.

18. See, for example, G. Popkin, "Update the Nobel Prizes," *New York Times*, October 3, 2016.

19. Original interview, Cambridge, MA, October 4, 2016.

20. Original interview, Cambridge, MA, October 4, 2016.

CHAPTER 2

1. See L. Gårding and L. V. Hörmandr, "Why Is There No Nobel Prize in Mathematics?," *Mathematical Intelligence* 7 (1985): 73–74.

2. See, for example, S. Widmalm, "Introduction to the Special Edition," *Perspectives on the Prize: Essays in Commemoration of the First Century of the Nobel Prizes*," *Minerva* 39 (2001): 365–372.

3. It is clear that this data can be related to the wider context of the dynamics that characterize the presence of women in scientific careers, particularly in the highest ranks of academia. See, for example, UNESCO Science Report 2021: Statistics and Resources, https://www.unesco.org/reports/science/2021/en/statistics; Eurostat, Human Resources in Science and Technology, https://ec.europa.eu/eurostat/statistics-explained.

4. R. Björk, "The Age at Which Nobel Prize Research Is Conducted," *Scientometrics* 119, no. 2 (2019): 913–939.

5. If these figures are used as a "performance indicator" for different countries, some say that they should be weighted on the basis of population to avoid improperly comparing much larger and more populated countries with much smaller and less populated ones, as is done with the number of researchers per workforce. This way,

countries such as Switzerland, Sweden, or the Netherlands take the lead with significant numbers of Nobel Prizes compared to the general population.

6. H. Zuckerman, *Scientific Elite: Nobel Laureates in the United States* (New York: Free Press, 1977).

7. L. S. Sherby, *The Who's Who of Nobel Prize Winners, 1901–2000*, 2nd ed. (Westport, CT: Oryx Press, 2002); B. Feldman, *The Nobel Prize: A History of Genius, Controversy, and Prestige* (New York: Arcade, 2000); I. Hargittai, *The Road to Stockholm: Nobel Prizes, Science, and Scientists* (Oxford: Oxford University Press, 2002). On the theme of the relationship between science and religion, especially in relation to modern science, see the classic R. K. Merton, *Scienza, religione e politica* (*Science, Religion, and Politics*) (Bologna: Il Mulino, 2011). See also L. Critelli, *Minoranze di eccellenza scientifica. L'ascesa di ebrei e di altre minoranze americane nella ricerca* (*Minorities in Scientific Excellence: The Rise of Jews and Other American Minorities in Research*) (Bolzano: Il Brennero–Der Brenner, 2001).

8. E. Garfield and M. Malin, "Can Nobel Prize Winners Be Predicted?," paper presented at the 135th Annual Meeting, American Association for the Advancement of Science, Dallas, Texas, December 26–31, 1968, 6.

9. R. van Noorden et al., "The Top 100 Papers: *Nature* Explores the Most-Cited Research of All Time," *Nature*, 514, no. 7524 (October 30, 2014): 550–553.

10. E. Crawford, *The Beginnings of the Nobel Institution: The Science Prizes, 1901–1915* (Cambridge: Cambridge University Press, 1984); Hargittai, *The Road to Stockholm*.

11. Quoted in J. Jenkin, *William and Lawrence Bragg, Father and Son* (Oxford: Oxford University Press, 2008), 330.

12. W. H. Bragg, "X-rays and Crystals," *Nature* 90, no. 2248 (November 28, 1912): 360.

13. "Nomination and Selection of Physics Laureates," The Nobel Prize, https://www.nobelprize.org/nomination/physics.

14. See S. Roach, "Two Nobel Prize Gold Medals Sell at Auctions by Two Firms Days Apart," *Coin World*, November 5, 2015; N. St. Fleur, "Why the Scientist Who Unravelled DNA Is Selling His Nobel Prize," *The Atlantic*, December 1, 2014; D. Crow, "James Watson to Sell Nobel Prize Medal," *Financial Times*, November 28, 2014.

CHAPTER 3

1. E. Crawford, *The Beginnings of the Nobel Institution: The Science Prizes, 1901–1915* (Cambridge: Cambridge University Press, 1984). Gullstrand's speech can be found on the prize's website at The Nobel Prize, https://www.nobelprize.org.

2. Crawford, *The Beginnings of the Nobel Institution*, 191.

3. Crawford, *The Beginnings of the Nobel Institution*, 191.

4. *The Times* (London), January 6, 7, 8, 1902: "Among the authors of letters were E. Gosse, librarian to the House of Lords and a well-known literary figure, and S. P. Thompson, a physicist and active nominator for the science prizes," in Crawford, *The Beginnings of the Nobel Institution*, 193.

5. "Deutschland in der Welt voran!," headline in *Die Welt am Montag, Rheinisch-Westfalische Zeitung* and others, December 11, 1905, quoted in Crawford, *The Beginnings of the Nobel Institution*, 191.

6. *Vossische Zeitung*, August 20, 1911, quoted in Crawford, *The Beginnings of the Nobel Institution*, 192.

7. *La Liberté*, November 15, 1903, quoted in Crawford, *The Beginnings of the Nobel Institution*, 195.

8. M. W. Browne, "Americans at Harvard and Purdue Win—German and Pakistani Cited," *New York Times*, October 16, 1979.

9. T. Carlyle, *On Heroes, Hero-Worship and the Heroic in History* (London: James Fraser, 1841), 114–115.

10. Headline in the *Korea Times*, quoted in C. K. Chekar and J. Kitzinger, "Science, Patriotism and Discourses of Nation and Culture: Reflections on the South Korean Stem Cell Breakthroughs and Scandals," *New Genetics and Society* 26 (2007): 289–307. Hwang Woo-suk was later involved in a scandal and forced to retract some publications based on falsified data.

11. Cong Cao, "Chinese Science and the 'Nobel Prize Complex,'" *Minerva* 42 (2004): 169.

12. Cao, "Chinese Science," 169.

13. Coubertin, quoted in A. Fugardi, *Storia delle Olimpiadi* (*History of the Olympics*) (Bologna: Cappelli, 1958), 70.

14. S. Widmalm, "Introduction," *Minerva* 33 (1995): 370.

15. B. Schroeder-Gudehus, *Les scientifiques et la paix. La communauté scientifique internationale au cours des années 20* (Montreal: Les Presses de l'Université de Montréal, 1978), 49, quoted in S. Widmalm, "Science and Neutrality: The Nobel Prizes of 1919 and Scientific Internationalism in Sweden," *Minerva* 33 (1995): 339.

16. Widmalm, quoted in "Science and Neutrality," 340.

17. For example, "The Olympics of the Mind," *Forbes*, September 26, 2013.

18. See A. Kruger and W. Murray, eds., *The Nazi Olympics: Sport, Politics, and Appeasement in the 1930s* (Urbana: University of Illinois Press, 2003).

19. EurekAlert!, American Association for the Advancement of Science, https://www.Eurekalert.org, September 2, 2008.

20. The three quotes come, respectively, from James Fallow, "The Significance of a Nobel Prize for a Chinese Scientist," *The Atlantic*, October 5, 2015, https://www.theatlantic.com; "For China, Nobel Prize in Science Is Still a Big Leap Away," *South China Morning Post*, October 18, 2013, https://www.scmp.com; Junbo Yu, "The Politics behind China's Quest for Nobel Prizes," *Issues in Science and Technology* 30 (Spring 2014), https://issues.org.

21. The translation is from A. Guerraggio, *La scienza in trincea. Gli scienziati italiani nella prima guerra mondiale* (*Science in the Trenches: Italian Scientists in the First World War*) (Milan: Raffaello Cortina, 2015), 93.

22. R. M. Friedman, *The Politics of Excellence: Behind the Nobel Prize in Science* (New York: Times Books, 2001), 78.

23. Quoted in Friedman, *The Politics of Excellence*, 77.

24. Quoted in J. Jenkin, *William and Laurence Bragg, Father and Son* (Oxford: Oxford University Press, 2008), 347.

25. Quoted in J. Jenkin, "A Unique Partnership: William and Lawrence Bragg and the 1915 Nobel Prize in Physics," *Minerva* 39 (2001): 389.

26. Friedman, *The Politics of Excellence*, 90.

27. S. Widmalm, "Science and Neutrality," 352.

28. Friedman, *The Politics of Excellence*, 113.

29. Quoted in Widmalm, "Science and Neutrality," 351.

30. *Corriere della Sera*, February 5, 1920, 2.

31. *Corriere della Sera*, June 3, 1920, 4.

32. Quoted in B. Feldman, *The Nobel Prize: A History of Genius, Controversy, and Prestige* (New York: Arcade Publishing, 2000), 234.

33. J. Bernstein, *Hitler's Uranium Club: The Secret Recordings at Farm Hall*, 2nd ed. (New York: Copernicus Books, 2001), 284.

34. For more on the fascinating history of the classification of the platypus, see S. J. Gould, *Bully for Brontosaurus: Reflections in Natural History* (New York: Norton, 1991); U. Eco, *Kant e l'ornitorinco (Kant and the Platypus)* (Milan: Bompiani, 1997).

35. See K. Gavroglu, "Appropriating the Atom at the End of the 19th Century: Chemists and Physicists at Each Other's Throat," in *Philosophers in the Laboratory*, ed. V. Mossini (Modena: National Academy of Sciences, Letters and Arts, 1995).

36. Crawford, *The Beginnings of the Nobel Institution*, 121.

37. D. Knight, *Ideas in Chemistry* (New Brunswick, NJ: Rutgers University Press, 1992), 11–12.

38. I. Hargittai, *The Road to Stockholm: Nobel Prizes, Science, and Scientists* (Oxford: Oxford University Press, 2002), 43.

39. On the Nobel Prize to Watson and Crick and in particular on the role of Rosalind Franklin in the rewarded discovery of the structure of DNA, see chapter 5.

40. M. Howorth, *The Life Story of Frederick Soddy* (London: New World Publications, 1958), 83.

CHAPTER 4

1. For the story of the Nobel Prize to Einstein, besides the original documents from the archives of the Royal Swedish Academy of Sciences, I used among others: A. Elzinga, *Einstein's Nobel Prize: A Glimpse behind Closed Doors. The Archival Evidence* (Sagamore Beach, MA: Science History Publications, 2006); R. M. Friedman, *The Politics of Excellence. Behind the Nobel Prize in Science* (New York: Times Books, 2001); P. Galison, *Einstein's Clocks, Poincaré's Maps: Empires of Time* (New York: W. W. Norton, 2003);

A. Pais, *Subtle Is the Lord: The Science and the Life of Albert Einstein* (Oxford: Oxford University Press, 1982); J. Renn, *Albert Einstein: Chief Engineer of the Universe* (Berlin: Wiley, 2005).

2. "On the Electrodynamics of Moving Bodies," *Annalen der Physik* 17 (1905): 891–921, in *The Collected Papers of Albert Einstein*, vol. 2, *The Swiss Years: Writings, 1900–1909*, English translation supplement, ed. J. Stachel et al. (Princeton, NJ: Princeton University Press, 1989), 141.

3. A. Einstein, "Does the Inertia of a Body Depend upon Its Energy-Content?," *Annalen der Physik* 18 (1905): 639–641, in *The Collected Papers of Albert Einstein*, vol. 2, *The Swiss Years: Writings, 1900–1909*, English translation supplement, ed. J. Stachel et al. (Princeton, NJ: Princeton University Press, 1989), 174.

4. Einstein, "Does the Inertia of a Body Depend upon Its Energy-Content?"

5. Quoted in M. Wazeck, "Einstein in the Daily Press: A Glimpse into the Gehrcke Papers," in *In the Shadow of the Relativity Revolution*, ed. J. Renn et al. (Berlin: Max Planck Institute for the History of Science, 2004), 74.

6. "Letter to Max Planck, 6 July 1922," in *The Collected Papers of Albert Einstein*, vol. 13, *The Berlin Years: Writings & Correspondence January 1922–March 1923*, English translation supplement, ed. D. K. Buchwald et al. (Princeton, NJ: Princeton University Press, 1989), 212.

7. E. Cavalli, *Il romanzo del Nobel nel racconto di un inviato* (The Romance of the Nobel as Told by a Special Envoy) (Rome: Rai-Eri, 2000), 105.

8. F. Becattini et al., "The Nobel Prize Delay," *Physics Today*, May 27, 2014, http://physicstoday.scitation.org. See B. F. Jones and B. A. Weinberg, "Age Dynamics in Scientific Creativity," *Proceedings of the National Academy of Sciences of the United States of America* 108 (2011): 18910–18914.

9. Becattini et al., "The Nobel Prize Delay," 2.

10. Data from UNESCO Science Report 2021: Statistics and Resources, https://www.unesco.org/reports/science/2021/en/statistics. See also D. E. Ayan et al., "How Many People in the World Do Research and Development?," *Global Policy* 14, no. 2 (May 2023): 270–287; Asghar Ghasemi et al. "Scientific Publishing in Biomedicine: A Brief History of Scientific Journals," *International Journal of Endocrinology and Metabolism* 21, no. 1 (January 2023): e131812. On this theme, see also M. Bucchi, *Science in Society. An Introduction to Social Studies of Science* (London: Routledge, 2004); M. Bucchi, "Visible Scientist," in *Encyclopedia of Science and Technology Communication*, ed. S. H. Priest (Thousand Oaks, CA: Sage, 2010), 923–933; M. Bucchi, "Norms, Competition, and Visibility in Contemporary Science: The Legacy of Robert K. Merton," *Journal of Classical Sociology* 15 (2015): 233–252.

11. Original interview, Stockholm, May 12, 2013.

12. Becattini et al., "The Nobel Prize Delay," 2.

13. E. Norrby, *Nobel Prizes and Life Sciences* (Singapore: World Scientific Publishing, 2010), 115.

14. On the Nobel to Wagner-Jauregg, see M. Whitrow, "Wagner-Jauregg and Fever Therapy," *Medical History* 34 (1990): 294–310; E. Brown, "Why Wagner-Jauregg Won the Nobel Prize for Discovering Malaria Therapy for General Paresis of the Insane," *History of Psychiatry* 11 (2000): 371–382. On the Nobel to Finsen, see A. Grzybowski and K. Pietrzak, "From Patient to Discoverer: Niels Ryberg Finsen (1860–1904), the Founder of Phototherapy in Dermatology," *Clinics in Dermatology* 30 (2010): 451–455. On Fibiger, see C. M. Stolt et al., "An Analysis of a Wrong Nobel Prize: Johannes Fibiger, 1926. A Study in the Nobel Archives," *Advances in Cancer Research* 91 (2004): 1–12.

15. *Corriere della Sera*, October 27, 1927.

16. The reply from the Nobel Foundation was published by the British newspaper *The Guardian*, August 2, 2004. On the Nobel to Moniz, see V. W. Swayze II, "Frontal Leukotomy and Related Psychosurgical Procedures in the Era before Antipsychotics (1935–1954): A Historical Overview," *American Journal of Psychiatry* 152 (1995): 505–515; B. Jansson, "Controversial Psychosurgery Resulted in a Nobel Prize," 1998, The Nobel Prize, https://www.nobelprize.org.

CHAPTER 5

1. Quoted in R. L. Sime, "Belated Recognition: Lise Meitner's Role in the Discovery of Fission," *Journal of Radioanalytical and Nuclear Chemistry* 142 (1990): 22.

2. L. Meitner, "Looking Back," *Bulletin of the Atomic Scientists* 20 (1964): 4.

3. R. M. Friedman, *The Politics of Excellence: Behind the Nobel Prize in Science* (New York: Times Books, 2001), 237.

4. R. L. Sime, *Lise Meitner: A Life in Physics* (Berkeley: University of California Press, 1996), 323.

5. E. Crawford et al., "A Nobel Tale of Postwar Injustice," *Physics Today* 50 (September 1997): 32.

6. Sime, *Lise Meitner*, 339.

7. *Corriere della Sera*, October 8, 2009.

8. J. Boyd, "Sleeping Sickness: The Castellani-Bruce Controversy," *Notes and Records of the Royal Society* 28 (1973): 93.

9. Boyd, "Sleeping Sickness," 94.

10. Boyd, "Sleeping Sickness," 95.

11. D. Butler, "Close but No Nobel: The Scientists Who Never Won," *Nature*, October 11, 2016, https://www.nature.com.

12. P. Levi, *Il Sistema periodico* (Turin: Einaudi, 1975); *The Periodic Table*, trans. Raymond Rosenthal (New York: Schocken Books, 1984), 6.

13. I. Hargittai, *The Road to Stockholm: Nobel Prizes, Science, and Scientists* (Oxford: Oxford University Press, 2002).

14. B. Feldman, *The Nobel Prize: A History of Genius, Controversy, and Prestige* (New York: Arcade Publishing, 2000), 256.

15. H. Bredekamp, *Theorie des Bildakts* (Berlin: Suhrkamp, 2010).

16. J. D. Watson, *The Double Helix: A Personal Account of the Discovery of the Structure of DNA* (New York: Atheneum, 1968).

17. Watson, *The Double Helix*, 143–144.

18. K. Nightingale, "Behind the Picture: Photo 51," interview with Brian Sutton, *Insight*, April 25, 2013.

19. E. J. Yoxen, *The Social Impact of Molecular Biology* (Cambridge: University of Cambridge, 1977), 347; J. Turney, *Frankenstein's Footsteps: Science, Genetics and Popular Culture* (New Haven, CT: Yale University Press, 1998). Also the Italian press substantially ignored the discovery at the time (see G. Caprara, *L'avventura della scienza* (*The Adventure of Science*) (Milan: Rizzoli, 2009).

20. Quoted in E. Norrby, *Nobel Prizes and Nature's Surprises* (Singapore: World Scientific Publishing, 2013), 324.

21. Norrby, *Nobel Prizes and Nature's Surprises*, 325.

22. See B. Fantini, "The Concept of Specificity and the Italian Contribution to the Discovery of the Malaria Transmission Cycle," *Parassitologia* 41 (1999): 39–47; E. Capanna, "Battista Grassi: A Zoologist for Malaria," *Contributions to Science* 3 (2006): 187–195.

23. See chapter 1 and the epilogue. See also M. Bucchi, *Per un pugno di idee. Storie di innovazioni che hanno cambiato la nostra vita* (*A Fistful of Ideas: Stories of Innovations That Changed Our Lives*) (Milan: Bompiani, 2016).

24. Quoted in A. Aldridge, *The Beatles Illustrated Lyrics* (London: Macdonald, 1969).

25. H. W. Kroto et al., "C_{60}: Buckminsterfullerene," *Nature* 318 (November 1985): 162–163.

26. Quoted in Hargittai, *The Road to Stockholm*, 241.

27. In the original, Михаил СеменовичЦвѣтъ:t, at times transcribed as Tsvett.

28. M. Tswett, "Physikalisch-Chemische Studien über das Chlorophyll. Die Adsorption," *Berichte der Deutschen botanischen Gesellschaft* 24 (1906) 322. See also L. S. Ettre and K. I. Sakodynskii, "M. S. Tswett and the Discovery of Chromatography," parts 1 and 2, *Chromatographia* 35 (1993): 223–231, 329–338; L. S. Ettre, "M. S. Tswett and the 1918 Nobel Prize in Chemistry," *Chromatographia* 42 (1996): 343–351.

CHAPTER 6

1. See J. Gregory and S. Miller, *Science in Public: Communication, Culture, and Credibility* (London: Plenum, 1998).

2. 21st Century King James Version.

3. R. K. Merton, "The Matthew Effect in Science," in *The Sociology of Science: Theoretical and Empirical Investigations* (Chicago: University of Chicago Press, 1973), 447.

4. Merton, "The Matthew Effect in Science," 443.

5. See J. Cole and S. R. Cole, *Social Stratification in Science* (Chicago: University of Chicago Press, 1973); H. Inhaber and K. Przednowek, "Quality of Research and the Nobel Prizes," *Social Studies of Science* 6 (1976): 33–50.

6. Merton, "The Matthew Effect in Science," 457.

7. Merton, "The Matthew Effect in Science," 442.

8. H. Zuckerman, *Scientific Elite: Nobel Laureates in the United States* (1977) (New Brunswick, NJ: Transaction, 1996), 236–237.

9. Quoted in I. Hargittai, *The Road to Stockholm: Nobel Prizes, Science, and Scientists* (Oxford: Oxford University Press, 2002), 204.

10. Original interview, Tokyo, November 4, 2016.

11. Aphorism for a friend, September 18, 1930, Einstein Archive 36–598; also quoted in B. Hoffman with H. Dukas, *Albert Einstein: Creator and Rebel* (New York: Viking, 1972), 24.

12. O. Glasser, *Wilhelm Conrad Röntgen and the Early History of the Röntgen Rays* (San Francisco: Norman, 1993).

13. M. Pavese Rubins, "Guglielmo Marconi and the Wireless Telegraphy in the Swedish Daily Press," in *A Wireless World: One Hundred Years since the Nobel Prize to Guglielmo Marconi*, ed. K. Grandin et al. (Stockholm: Center for the History of Science at the Royal Swedish Academy of Sciences, 2012), 104.

14. *Corriere della Sera*, July 4–5, 1897, 1–2.

15. *Corriere della Sera*, July 4–5, 1897, 1–2.

16. Marconi and Holman never actually married and split soon after. Marconi eventually married Beatrice O'Brien in 1905.

17. See, for example, A. Henderson, "Media and the Rise of Celebrity Culture," *OAH Magazine of History* 6 (1992): 49–54.

18. E. Curie, *Madame Curie* (Paris: Gallimard, 1938).

19. On how the awarding of the Nobel Prize has sometimes been presented or perceived in contrast with the poor recognition of the winners in their own country, see chapter 3.

20. *Corriere della Sera*, December 12, 1903.

21. A. Pizzorno, "Dalla reputazione alla visibilita" ("From Reputation to Visibility"), in *Il velo della diversità. Studi su razionalità e riconoscimento (The Veil of Diversity: Studies on Rationality and Recognition)*, ed. A. Pizzorno (Milan: Feltrinelli, 2007), 229.

22. See H. M. Collins and T. Pinch, *The Golem: What Everyone Should Know about Science* (New York: Cambridge University Press, 1993); M. Bucchi, *Science in Society: An Introduction to Social Studies of Science* (New York: Routledge, 2004).

23. Pizzorno, *Dalla reputazione alla visibilità (From Reputation to Visibility)*, 323.

24. See B. V. Lewenstein, "Cold Fusion and Hot History," *Osiris*, 2nd ser., 7 (1992): 135–163; M. Bucchi, *Science and the Media: Alternative Routes in Scientific Communication* (London: Routledge, 1998).

25. R. Goodell, *The Visible Scientists* (Boston: Little, Brown, 1977), 4.

26. See H. P. Peters, "Scientists as Public Experts," in *Handbook of Public Communication of Science and Technology*, 2nd ed., ed. M. Bucchi and B. Trench (London: Routledge, 2014); M. Bucchi, "Visible Scientist," in *Encyclopaedia of Science and Technology Communication* (Thousand Oaks, CA: Sage, 2010), 932–933.

27. See chapter 5. Similarly, it emerges from data on public opinion such as the Osservatorio Scienza Tecnologia e Societa (Science and Technology in Society Monitor) that many interviewers tend to identify particularly visible scientists with the Nobel although they might have not received it and, vice versa, not to identify less familiar scientists as prize winners. See later in this chapter.

28. Cf. Asghar Ghasemi et al., "Scientific Publishing in Biomedicine: A Brief History of Scientific Journals," *International Journal of Endocrinology and Metabolism* 21, no. 1 (January 2023): e131812.

29. On these themes, see M. Bucchi, "Norms, Competition and Visibility in Contemporary Science: The Legacy of Robert K. Merton," *Journal of Classical Sociology* 15 (2015): 233–252; M. Bucchi and B. Trench, "Science Communication and Science in Society: A Conceptual Review in Ten Keywords," *Tecnoscienza* 7, no. 2 (2016): 151–168; D. Fahy and B. Lewenstein, "Scientists in Popular Culture," in *Handbook of Public Communication of Science and Technology*, new ed., ed. M. Bucchi and B. Trench (London: Routledge, 2014), 83–96; S. Rödder et al., eds., *The Sciences' Media Connection: Public Communication and Its Repercussions* (Dordrecht, Netherlands: Springer, 2012).

30. L. Beltrame, "Ipse dixit. I premi Nobel come argomento di autorità nella comunicazione pubblica della scienza" ("Ipse Dixit: The Nobel Prizes as a Voice of Authority in the Public Communication of Science"), *Studi di Sociologia* 45 (2007): 77–98.

31. See M. Mulkay and N. Gilbert, "Joking Apart: Some Recommendations Concerning the Analysis of Scientific Culture," *Social Studies of Science* 12 (1982): 585–614.

32. It is no coincidence that Braschi-Dulbecco is addressing Sister Germana, a nun known to the public for her successful cookbooks. The contrast between scientific activity—in particular, that taking place in a laboratory—with culinary practices is a common rhetorical strategy for making science more "human" by bringing it closer to common sense. M. Bucchi, *Newton's Chicken: Science in the Kitchen* (Singapore: World Scientific, 2020).

33. On this topic, see the classic by E. Goffman, *The Presentation of Self in Everyday Life* (New York: Doubleday, 1959); in relation to science communication, see Bucchi, *Science and the Media*.

34. For a more detailed analysis of the results of this study, see M. Bucchi and F. Neresini, "Un Nobel a Sanremo (ma la scienza rimane sconosciuta)" ("A Nobel in Sanremo—but the Science Remains Unknown"), *Problemi dell'informazione (Issues in Information)* 25 (2000): 233–250.

35. Hargittai, *The Road to Stockholm*, 65.

36. Translator's note: the main Italian encyclopedia.

37. M. Bucchi and B. Saracino, "'Visual Science Literacy': Images and Public Understanding of Science in the Digital Age," *Science Communication* 38 (2016): 812–819; M. Bucchi and B. Saracino, "Scienza, tecnologia e opinione pubblica in Italia nel 2016" ("Science, Technology, and Public Opinion in Italy in 2016"), in *Annuario scienza tecnologia e società 2017*, ed. B. Saracino (Bologna: Il Mulino, 2017).

38. Hargittai, *The Road to Stockholm*.

39. Eponymy sometimes applies to what Merton refers to as "obliteration by incorporation": some scientific contributions become such classics and taken for granted that their origin is forgotten. R. K. Merton, *Social Theory and Social Structure* (1949) (New York: Free Press, 1968). Ultimately, it is possible to be so well known and familiar as to be forgotten! On Golgi and the Nobel Prize, see P. Mazzarello, *Il Nobel dimenticato. La vita e la scienza di Camillo Golgi (The Forgotten Nobel: Life and Science of Camillo Golgi)* (Turin: Boringhieri, 2006), where, among others, a case is cited of medical students who believe the term "Golgi" (for example, as in the "Golgi apparatus") is derived from Latin.

CHAPTER 7

1. S. Shapin, *The Scientific Life: A Moral History of a Late Modern Vocation* (Chicago: University of Chicago Press, 2008).

2. J. Browne, "I Could Have Retched All Night: Charles Darwin and His Body," in *Science Incarnate: Historical Embodiments of Natural Knowledge*, ed. C. Lawrence and S. Shapin (Chicago: University of Chicago Press, 1998), 279.

3. S. Shapin, "The Philosopher and the Chicken," in Lawrence and Shapin, *Science Incarnate*, 43–45.

4. L. T. More, *Isaac Newton: A Biography* (1934) and R. S. Westfall, *Never at Rest: A Biography of Isaac Newton* (1980), cited in Shapin and Lawrence, *Science Incarnate*; P. De Kruif, *Microbe Hunters* (New York: Harcourt and Brace, 1926); P. Camporesi, *Le officine dei sensi (The Workshops of the Senses)* (1985) (Milan: Garzanti, 2009). On this theme, see M. Bucchi, *Newton's Chicken: Science in the Kitchen* (Singapore: World Scientific, 2020).

5. Shapin, "The Philosopher and the Chicken," 36.

6. The quotes come, respectively, from Conforti, in M. Beretta et al., eds., *Savant Relics: Brains and Remains of Scientists* (Sagamore Beach, MA: Science History Publications, 2016), viii; E. Irace, *Itale glorie* (Bologna: Il Mulino, 2003), 87. On how the cult of saints in its turn had built upon the cult of pagan heroes, see, for example, M. T. Fumagalli Beonio Brocchieri and G. Guidorizzi, *Corpi gloriosi. Eroi greci e santi cristiani (Glorious Bodies: Greek Heroes and Christian Saints)* (Rome: Laterza, 2012).

7. T. Carlyle, *On Heroes: Hero-Worship and the Heroic in History* (London: James Fraser, 1841), 133.

8. M. Shortland and R. Yeo, *Telling Lives in Science: Essays on Scientific Biography* (Cambridge: Cambridge University Press, 1996), 113.

9. G. Cantor, "Where the Statue Stood: Celebrations of Faraday 1867–1931," paper presented at the Royal Institution conference on Science and Its Publics, London, September 22, 1994, 8.

10. Another fascinating and significant case is that of the monument erected in memory of the philosopher Giordano Bruno in the Campo de' Fiori in 1889. See M. Bucciantini, *Campo dei Fiori. Storia di un monumento maledetto* (*Campo dei Fiori: History of a Cursed Monument*) (Turin: Einaudi, 2015).

11. L. Moschen, "Discorso per l'inaugurazione del monumento a Canestrini a Trento" ("Speech for the Inauguration of the Monument to Canestrini in Trento"), *Tridentum* (Trento) 5 (1902): 338.

12. The bust was later defaced by unknown persons and then replaced with a bronze copy, and the marble original was moved to the premises of the Tridentine Museum of the Natural Sciences (currently the MUSE), where it is still displayed.

13. *Alto Adige*, September 15–16, 1902, 2.

14. G. L. Geison, *The Private Science of Louis Pasteur* (Princeton, NJ: Princeton University Press, 1995), 263.

15. François Jacob, 1988, quoted in Geison, *The Private Science of Louis Pasteur*, 263.

16. E. H. Kantorowicz, *The King's Two Bodies: A Study in Mediaeval Political Theology* (Princeton, NJ: Princeton University Press, 1957).

17. Geison, *The Private Science of Louis Pasteur*, 271.

18. F. Zampieri et al., "Nota storico-medica sulla quinta vertebra lombare di Galileo Galilei conservata presso l'Università di Padova" ("Historic-Medical Note on the Fifth Lumbar Vertebra of Galileo Galilei Preserved at the University of Padova"), in *I Medici. Uomini, potere e passione* (*The Medici: Men, Power, and Passion*), ed. A. Wieczorek et al. (Regensburg, Germany: Schnell und Steiner, 2013), 348–351; A. Zanatta et al., "New Interpretation of Galileo's Arthritis and Blindness," *Advances in Anthropology* 5 (2015): 39–49. For other richly documented examples and on the general theme the relics of scientists, see Beretta et al., *Savant Relics*.

19. A. Castronuovo, *Ossa, cervelli, mummie e capelli* (*Bones, Brains, Mummies, and Hair*) (Macerata: Quolibet, 2016), 45.

20. M. Paterniti, *Driving Mr. Albert: A Trip across America with Einstein's Brain* (New York: Dial Press, 2000).

21. U. Larsson, ed., *Cultures of Creativity: The Centennial Exhibition of the Nobel Prize* (Canton, MA: Science History Publications and the Nobel Museum, 2001).

22. S. Widmalm, "Introduction to the Special Issue *Perspectives on the Prize: Essays in Commemoration of the First Century of the Nobel Prizes*," *Minerva* 39 (2001): 365–372; M. Bloch, *Les rois thaumaturges* (Paris: Max Leclerc, 1961); English translation: *The Royal Touch* (London: Routledge, 1973).

23. K. Woodward, *Making Saints: How the Catholic Church Determines Who Becomes a Saint, Who Doesn't, and Why* (New York: Simon and Schuster, 1990).

24. Woodward, *Making Saints*, 78.

25. Quoted in I. Hargittai, *The Road to Stockholm: Nobel Prizes, Science, and Scientists* (Oxford: Oxford University Press, 2002), 57.

26. E. Regis, *Who Got Einstein's Office? Eccentricity and Genius at the Institute for Advanced Study* (Reading, MA: Addison-Wesley, 1987).

27. Regis, *Who Got Einstein's Office?*, 32.

28. See, for example, BabyCenter, https://www.babycenter.com.

29. E. Goffman, "The Nature of Deference and Demeanor," *American Anthropologist* 58, no. 3 (June 1956): 473–502.

30. Goffman, "The Nature of Deference and Demeanor," 477.

31. John Polanyi, speech at the Nobel banquet, December 10, 1986, The Nobel Prize, https://www.nobelprize.org.

32. D. Dayan and E. Katz, *Media Event: The Live Broadcasting of History* (Cambridge, MA: Harvard University Press, 1992).

33. Y. Zotterman, *Touch, Tickle and Pain* (Oxford: Pergamon Press, 1971), 266.

34. B. Feldman, *The Nobel Prize: A History of Genius, Controversy, and Prestige* (New York: Arcade Publishing, 2000), 275.

35. Goffman, "The Nature of Deference and Demeanor," 489.

36. The expression is used by Isaac Newton in a letter to Robert Hooke dated February 5, 1676. Merton traces its origins back to Bernard of Chartres in the twelfth century. R. K. Merton, *On the Shoulders of Giants: A Shandean Postscript* (1965) (Orlando, FL: Harcourt Brace, 1985).

37. *Corriere della Sera*, October 27, 1959.

38. Richet, 1927, quoted in R. K. Merton, *The Sociology of Science: Theoretical and Empirical Investigations*, ed. N. W. Storer (Chicago: University of Chicago Press, 1973), 304.

39. M. Mulkay, "Noblesse Oblige," in *Sociology of Science: A Sociological Pilgrimage* (Bloomington: Indiana University Press, 1991), 169–182. This aspect is also characterized by the "coronation" as a media event. "Coronation [. . .] puts a great man on view [. . .] acting a ritual role, in which initiative in minimized, dwarfed by the august setting . . ." (D. Dayan and E. Katz, *Changing Conceptions of Leadership* [New York: Springer, 1986], 140).

40. *Corriere della Sera*, December 12, 1903.

41. D. Crow, "James Watson to Sell Nobel Prize Medal," *Financial Times*, November 28, 2014.

42. The definition of *nonperson* comes from the classic by E. Goffman, *The Presentation of Self in Everyday Life* (New York: Doubleday, 1959).

43. See chapter 2. For Watson's statements, see, for example, Crow, "James Watson to Sell Nobel Prize Medal."

44. Merton, *The Sociology of Science*, 276.

45. Merton, *The Sociology of Science*, 276.

46. The Boyle quote comes from G. Burnet, *A Funeral Sermon Preached at the Funeral of the Honourable Robert Boyle* (London, 1692), 25; the Bacon note comes from F. Bacon, *The Advancement of Learning* (1605).

47. A. C. Crombie, "Experimental Science and the Rational Artist in Early Modern Europe," *Daedalus* 116 (1986): 49–74, reprinted in A. C. Crombie, *Science, Art and Nature in Medieval and Modern Thought* (London: Hambledon, 1996), 90.

48. A. Smith, *Theory of Moral Sentiments* (1759), 6th ed. (London: Millar, 1790), 112.

49. Quoted in Shapin, *The Scientific Life*, 24.

50. Shapin, *The Scientific Life*, 79.

51. Both quotes from D. Pratt, *Nobel Wisdom: The 1000 Wisest Things Ever Said* (London: JR Books, 2008), 9, 143–144.

52. On the distinction between front-stage and back-stage in social interaction, see the classic Goffman, *The Presentation of Self in Everyday Life*; in relation to science communication, see M. Bucchi, *Science and the Media: Alternative Routes in Scientific Communication* (London: Routledge, 1998).

53. Merton, *The Sociology of Science*, 326.

54. J. Monod, *Le hasard et la nécessité* (Paris: Ed. Du Seuil, 1970); English translation: *Chance and Necessity: An Essay on the Natural Philosophy of Modern Biology* (New York: Vintage, 1972), 17.

55. Wilhelm Solf's report cited in Thomas F. Glick, *Einstein in Spain: Relativity and the Recovery of Science* (Princeton, NJ: Princeton University Press, 1988), 97–98.

56. Tsutomu Kaneko, "Einstein's Impact on Japanese Intellectuals," in *The Comparative Reception of Relativity*, ed. Thomas F. Glick (Dordrecht: Reidel, 1987), 362.

57. *Corriere della Sera*, December 11, 1986.

58. R. Levi-Montalcini, *Elogio dell'imperfezione* (*In Praise of Imperfection*) (Milan: Garzanti, 1990).

59. On the Repository for Germinal Choice, the official name of the initiative founded by millionaire Robert Graham, see the various articles published in *Slate* magazine in February 2001 at https://www.slate.com.

60. Hargittai, *The Road to Stockholm*, 69.

61. R. Jacques, "7 Unbelievable and Hilarious Ways Nobel Prize Winners Found Out They Had Won," HuffPost, Huffingtonpost.com, October 10, 2013.

62. Jacques, "7 Unbelievable and Hilarious Ways."

63. Milton Friedman, "The Nobel Prize in Economics, 1976," remarks delivered at the Income Distribution Conference, Hoover Institution, Stanford University, January 29, 1977, https://miltonfriedman.hoover.org/internal/media/dispatcher/215528/full.

64. P. Wilhelm, *The Nobel Prize* (London: Springwood Books, 1983), 65.

65. *Corriere della Sera*, October 28, 1959.

66. *Corriere della Sera*, October 28, 1959.

67. Helen Grady, "Profile: Andre Geim," BBC Radio 4, first broadcast July 27, 2013, https://www.bbc.co.uk/programmes/b037ghx1.

68. On the distance from the role, see E. Goffman, *Encounters: Two Studies in the Sociology of Interaction* (Indianapolis: Bobbs-Merrill, 1961). The Rubbia quote is in Pratt, *Nobel Wisdom*, 10.

69. Goffman, *Encounters*.

70. K. Mullis, *Dancing Naked in the Mind Field* (New York: Pantheon, 1998).

71. Jacques, "7 Unbelievable and Hilarious Ways."

EPILOGUE

1. T. Veblen, "The Place of Science in Modern Civilization," *American Journal of Sociology* 11 (March 1906): 585–609.

2. S. Shapin, *The Scientific Life: A Moral History of a Late Modern Vocation* (Chicago: University of Chicago Press, 2008), 170. The term *serendipity* recalls the accidental nature of various scientific discoveries. See R. K. Merton and E. Barber, *The Travels and Adventures of Serendipity* (Princeton, NJ: Princeton University Press, 2006). Frequently cited examples are Fleming's discovery of penicillin and the discovery of X-rays by Röntgen.

3. On role distance, see E. Goffman, *Encounters: Two Studies in the Sociology of Interaction* (Indianapolis: Bobbs-Merrill, 1961); regarding the intersection of social circles, see G. Simmel, *Die Kreuzung sozialer Kreise*, in G. Simmel, *Georg Simmel Gesamtausgabe* (1908), vol. 11, ed. O. Rammstedt (Frankfurt am Main: Suhrkamp, 1992), 456–511; on the social roles of scientists, see F. Znaniecki, *The Social Role of the Man of Knowledge* (New York: Columbia University Press, 1968); R. K. Merton, *The Sociology of Science: Theoretical and Empirical Investigations*, ed. N. W. Storer (Chicago: University of Chicago Press, 1973).

4. See, for example, A. Posner, *Public Intellectuals: A Study of Decline*, new ed. (Cambridge, MA: Harvard University Press, 2003). Another factor to bear in mind is that other figures may have now occupied some of the space that is traditionally occupied by Nobel Prize winners: for example, the mythologies built in recent years around the protagonists of the digital age (Steve Jobs, Bill Gates, Mark Zuckerberg, Elon Musk), celebrities that are accorded influence and authority on wide-ranging political, social, and economic topics.

5. John Polanyi, speech at Nobel banquet, December 10, 1986, The Nobel Prize, https://www.nobelprize.org.

INDEX

Académie des sciences, 10, 21
Académie Française, 104
Academy of Sciences, Royal Swedish, 8, 27–28, 42, 66
African trypanosomiasis, 87–88
Agre, Peter, 116, 144
Altman, Sidney, 54
Alvarez, Luis, 148
Amano, Hiroshi, 105
Anderson, Carl, 96
Antibacterial therapy, 49
Appleton, Edward, 55
Arrhenius, Svante, 43–44, 47, 51–52, 55, 65
Artemisia annua, 38
Ashkin, Arthur, 26
Astrophysics, 55
Austria, 32
Avery, Oswald, 54, 91, 94–95

Bacon, Francis, 147
Bacteriology, 54
Baeyer, Adolf von, 36, 41
Baillet, Adrien, 136
Balling, Rudi, 41

Baltimore, David, 149
Bank of Sweden, 18
Banting, Frederick, 23–24, 143, 152–153
Bárány, Anders, 97
Bárány, Robert, 43, 71, 151
Bardeen, John, 125, 144, 152–153
Barkla, Charles Glover, 44, 47, 109
Barthes, Roland, 150
Bartlett, Neil, 90–91
BASF, 45
Beadle, George, 152
Becquerel, Henri, 51–52, 115, 126
Behring, Emil von, 10, 89
Békésy, Georg von, 32
Bell, Jocelyn, 96–97, 100
Bell Laboratories, 21
Berg, Paul, 54, 125
Berlin, 40–41
Bernal, John Desmond, 92
Best, Charles, 23–24
Bethe, Hans, 55
Bichat, Xavier, 135
Biochemistry, 53, 54
Blackett, Patrick, 20, 96
Blobel, Günter, 32

Bloch, Arthur, 55
Block, Marc, 141
Bohr, Niels, 15, 20, 27, 33, 79–83
Born, Max, 149
Bosch, Carl, 45–46
Boström, Erik Gustaf, 10
Bottaro-Costa, Francesco, 110
Boyle, Robert, 147
Bragg, Robert, 43
Bragg, William Henry, 21, 26–27, 43, 47, 71, 87, 109
Bragg, William Lawrence, 21, 26–27, 43, 47, 53, 71, 87, 95, 109, 152–153
Bragg's law, 26
Branting, Hjalmar, 6–7, 12, 46
Brattain, Walter, 125
Braun, Karl Ferdinand, 85
Breakthrough Prizes, 15–16
Bredekamp, Horst, 92
Brown, Louise, 72
Bruce, David, 88
Butenandt, Adolf Friedrich Johann, 48–49, 67

Calculators, 22
California Institute of Technology, 19
Calmette, Albert, 90
Cambridge University, 19
Canestrini, Giovanni, 136–137
Canonization ceremonies, 141
Carlsson, Arvid, 155
Carl XVI Gustav, king of Sweden, 31, 151
Carlyle, Thomas, 37, 136
Carrel, Alexis, 55, 131, 145
Casta, Laetitia, 120–122, 124
Castellani, Aldo, 88–89, 91
Cavendish Institute, 53
Cech, Thomas, 54
Celebrity scientists, 104–105, 114, 116, 118–119
Chadwick, James, 20
Chalfie, Martin, 157

Chance and Necessity, 131, 150
Chandrasekhar, Subrahmanyan, 96
Chemical physics, 51–52
Chemistry, Nobel Prize for
  Agre, Peter, 116
  Altman, Sidney, 54
  Arrhenius, Svante, 43
  Baeyer, Adolf von, 36
  Berg, Paul, 54
  Butenandt, Adolf Friedrich Johann, 48, 67
  Cech, Thomas, 54
  Chalfie, Martin, 157
  Curie, Marie, 51
  Curl, Robert, Jr., 99
  de Hevesy, George, 33
  Euler-Chelpin, Hans von, 48
  Fenn, John B., 26
  Gajdusek, Carleton, 145
  Gilbert, Walter, 54
  Haber, Fritz, 46
  Hahn, Otto, 49, 51
  Kendrew, John, 95
  Kobilka, Brian, 157
  Kroto, Harold, 40
  Kuhn, Richard, 48
  Martin, Archer J. P., 102
  Mullis, Kary, 20, 105
  Natta, Giulio, 143
  Northrop, John H., 94
  Ostwald, Wilhelm, 41
  Perutz, Max, 95
  Polanyi, John, 142
  Pregl, Fritz, 32
  Ramsay, William, 42
  Richards, Theodore W., 42
  Rutherford, Ernest, 56
  Ruzicka, Leopold, 48
  Sanger, Frederick, 54
  Semenov, Nikolaj, 53
  Sharpless, Barry, 125
  Smalley, Richard, 99
  Smith, Michael, 54

Synge, Richard L. M., 102
Tiselius, Arne, 53
van 't Hoff, Jacobus Henricus, 10
Virtanen, Artturi Ilmari, 50
Willstätter, Richard Martin, 43
Woodward, Robert, 141
China, 38–39, 41
Chlorine gas, 45
Chromatography, 101–102
Chronometers, 59
Citation counts, 20
Cockcroft, John, 20
Cohen, Stanley, 128
Cold fusion, 117
Collip, Bertram, 24
Comte, Auguste, 135
Conduitt, John, 136
Cooper, Leon, 125
Copley medal, 21
Cornell, Eric, 144
Coster, Dirk, 80
Coubertin, Pierre de, 39–40, 48
Crick, Francis, 33–34, 53–54, 92, 94–95, 99, 105, 126, 131, 149, 154
Curie, Marie, 13, 15, 25, 27, 32, 51–52, 105, 115–116, 126, 130–131, 145
Curie, Pierre, 51–52, 115–116, 145
Curl, Robert, Jr., 99, 103, 126

Dalén, Nils Gustaf, 75, 87
Damadian, Raymond Vahan, 97
Danish laureates, 129
Darwin, Charles, 135–136
Davis, Raymond, Jr., 25–26
Davy medal, 21
Deference, 142–143
De Hevesy, George, 33, 140
Demeanor, 143–144
Descartes, René, 104, 136
Diabetes, 23
Diatomite, 2
Diderot, Denis, 104
Dirac, Paul, 27

"Disintegration of Uranium by Neutrons: A New Type of Nuclear Reaction," 81
DNA, 54, 91–94, 131
Dolly the sheep, 123
Domagk, Gerhard, 48–49, 67, 152
The Double Helix, 131, 149–150
Driving Mr. Albert: A Trip across America with Einstein's Brain, 140
Dulbecco, Renato, 120–126, 129, 131
Dylan, Bob, 67
Dynamite, 2

Echegaray, José, 13
Economic sciences, prize in, 18
Eddington, Arthur, 55
Edison, Thomas Alva, 13, 85–87, 111, 118
Edwards, Robert, 72
Ehrenborg, Sigurd, 4
Ehrlich, Paul, 41
Einstein, Albert, 15, 27, 32, 44, 57, 59–66, 71, 75, 81, 87, 106, 117, 126, 129–131, 133–134, 139–140, 142, 150–151
Einstein, Mileva (nee Marić), 32, 60, 65
Enders, John, 91
Englert, François, 15
Euler-Chelpin, Hans von, 48, 50
Exner, Franz, 107

Faraday, Michael, 136
Fazio, Fabio, 120, 122, 124–125
Fenn, John B., 26
Fermi, Enrico, 25, 27, 51, 126, 141
Feynman, Richard, 32, 157
Fibiger, Johannes, 75
Finsen, Niels, 75
Fission, 81, 83
Flaubert, Gustave, 104
Forssell, Hans, 8
Fourth persons, 91, 95, 99–100
Fowler, William A., 96

Franck, James, 33, 47, 82
Franklin, Rosalind, 93, 95, 99
Freeman, Walter, 77
French Academy of Sciences, 10, 21
French prize-winners, 37
Freud, Sigmund, 91
Friedman, Milton, 154
Frisch, Robert, 80–81, 85
Fuller, Buckminster, 131
Fullerene, 131

Gadelius, Bror, 76
Gajdusek, Carleton, 145
Galileo Galilei, 136, 139, 147
Galvani, Luigi, 136
Gamow, George, 157
Garfield, Eugene, 20
Gehrcke, Ernst, 64
Geim, Andre, 156
Genetics, 54
Genius, scientist as, 160
Geophysics, 55
Germanium, 22
German prize-winners, 19, 36
Germany
  hostility to relativity, 62, 66
  Manifesto of the 93, 41–42, 45
  Nazi boycott, 48–49
  and Swedish scientists, 44, 84
Ghez, Andrea Mia, 25
Giacconi, Riccardo, 126, 132
Giacomini, Carlo, 139
Gilbert, Walter, 54, 125
Glace Nobel, 34
Goeppert-Mayer, Maria, 24–25, 141
Goffman, E., 146
Golgi, Camillo, 129
Golgi apparatus, 131
Goodell, R., 117
Gosling, Raymond, 93
Gotthard Tunnel, 2
Grand Hotel, 34
Grandin, Karl, 73

Grassi, Giovanni Battista, 95–96
Gravitational waves, 16, 116–117
Greengard, Paul, 32
Greenhouse effect, 51
Gregory, Richard, 148
Grossmann, Marcel, 59
Gullstrand, Allvar, 36, 61–65
Gustav, prince of Sweden, 10
Gustav V, king of Sweden, 65, 143

Haber, Fritz, 45–48, 101
Haber-Bosch process, 45
Hack, Margherita, 130
Hahn, Otto, 47, 49, 51, 71, 79–85
Haller, Friedrich, 59
Hammarskjöld, Dag, 18
Haroche, Serge, 32
Harrison, Frederic, 135
Harrison, George, 100
Harvard University, 19
Harvey, Thomas, 139–140
Heidenstam, Verner von, 11, 141
Heisenberg, Werner, 27, 82
Heldin, Carl-Henrik, 40
Hemingway, Ernest, 67
Hertz, Gustav, 47
Hess, Sofie, 4, 7
Hess, Victor, 55
Hevesy, George de, 33, 140
Hewish, Antony, 96
Higgs, Peter, 15, 74, 153
Higgs' boson, 15
Hiroshima, 82
Hitler, Adolf, 47, 80
Hodgkin, Alan Lloyd, 33
Hoff, Jacobus Henricus van 't, 10
Holman, Josephine, 148
Hoyle, Fred, 96
Hubble, Edwin, 55
Huffman, Donald, 99
Hunt, Tim, 128
Hwang, Woo-suk, 37
Hwass, Leonard, 4

Ig Nobel Prizes, 155–156
Immerwahr, Clara, 45
Inert gases, 90
Ingold, Christophe, 90
Insulin, 24
Integrated circuits, 22
In vitro fertilization (IVF), 72
Irony, 156–157
Islets of Langerhans, 23
Italian laureates, 128–130, 132

Jacob, François, 138, 157
Japan, 37
Jensen, Hans, 25
Jewish prize winners, 20
Johnson, Edward H., 86

Kaiser, David, 15
Kaiser Wilhelm Institut, 79
Kantorowicz, E. H., 138
Kao, Charles Kuen, 38
Kapitsa, Piotr, 20
Karlfeldt, Erik Axel, 18
Karolinska Institute, 8, 27–29, 42–43, 54
Kendrew, John, 53–54, 95
Ketterle, Wolfgang, 32
Kilby, Jack, 21–23, 25, 72
Kinsky von Wchinitz und Tettau, Bertha, 3–4, 12
Kobilka, Brian, 157
Koch, Robert, 36, 96
Kocher, Emil Theodor, 55
Korea, 37
Krätschmer, Wolfgang, 99
Krebs, Hans, 20
Krebs cycle, 20
Kroto, Harold, 40, 103, 126, 131
Kuhn, Richard, 48–49, 67, 145–146

Lagerlöf, Selma, 11
Landau, Lev, 71
Langevin, Paul, 142
Lattes, César, 96

Laue, Max von, 26–27, 33, 42, 50, 109
Lauterbur, Paul, 97
Lawrence, Ernest Orlando, 67
Lay Down Your Arms, 4
Lecher, Ernst, 107
Lee, Tsung Dao, 38
Lenard, Philipp von, 36, 41, 44, 62, 64, 108
Lennon, John, 100
Leriche, René, 89
Letterstedt Prize, 2
Leucotomy, 77
Levi, Primo, 90–91
Levi-Montalcini, Rita, 105, 119, 121, 124, 126, 128–129, 131, 151–153
L'Huillier, Anne, 25
Light, study of, 61–62
Light-emitting diode (LED), 105
LIGO experiment, 16
Liljestrand, Göran, 82
Lilljequist, Rudolf, 6–8
Lindegren, Agi, 10
Lippmann, Gabriel, 75
Literature, Nobel Prize for, 12
  Dylan, Bob, 67
  Heidenstam, Verner von, 141
  Prudhomme, Sully, 10
Lobotomy, 76–77
Loeb, Jacques, 90
Lombroso, Cesare, 137
Low, G. Carmichael, 88
Lowry, Oliver, 20
Luria, Salvatore, 105

Macleod, John, 23–24
Malaria, 38, 76, 95
Man, the Unknown, 131
Manifesto of the 93, 41, 45
Mansfield, Peter, 97
Marconi, Guglielmo, 13, 27, 55, 85, 89, 105, 109–114, 116, 130, 148
Margaret, princess of Sweden, 47
Marshall, Barry, 152–153

Martin, Archer J. P., 102
Massachusetts Institute of Technology (MIT), 19
Mathematics, exclusion of, 17
Matthew effect, 103–104, 117–118, 126, 162
Mayer, Joseph, 24
McCartney, Paul, 100
McClintock, Barbara, 21
Medicine, Nobel Prize for
  Baltimore, David, 149
  Banting, Frederick, 24
  Bárány, Robert, 43
  Beadle, George, 152
  Blobel, Günter, 32
  Carlsson, Arvid, 155
  Carrel, Alexis, 55
  Cohen, Stanley, 128
  Crick, Francis, 33
  Domagk, Gerhard, 48
  Dulbecco, Renato, 120
  Edwards, Robert, 72
  Ehrlich, Paul, 41
  Finsen, Niels, 75
  Greengard, Paul, 32
  Gullstrand, Allvar, 61–62
  Hodgkin, Alan Lloyd, 33
  Jacob, François, 157
  Koch, Robert, 36
  Kocher, Emil Theodor, 55
  Krebs, Hans, 20
  Levi-Montalcini, Rita, 128
  Luria, Salvatore, 105
  Macleod, John, 24
  Marshall, Barry, 153
  McClintock, Barbara, 21
  Moniz, Egas, 76
  Monod, Jacques, 150
  number of, 20
  Nurse, Paul, 32
  Nüsslein-Volhard, Christiane, 32
  Peyton Rous, Francis, 26
  Roberts, Richard, 32
  Ross, Ronald, 95
  Sanger, Frederick, 27
  Steinman, Ralph M., 29
  Tu Youyou, 38
  von Behring, Emil, 10
  von Békésy, Georg, 32
  Watson, James, 27
  Wilkins, Maurice, 54
Meer, Simon van der, 72, 128
Meitner, Lise, 49–50, 79–85, 99
Menkin, Miriam, 72
Merton, Robert K., 103–104, 118, 146–147
Michelson, Albert Abraham, 36–37
Microchips, 22
Millikan, Robert, 105
Milner, Yuri, 15
Mittag-Leffler, Gösta, 17
Modesty, 144
Moniz, Egas, 76–77, 145
Monod, Jacques, 131, 150
Moschen, Lamberto, 137
Moseley, Henry, 20, 42
Mullis, Kary, 20, 54, 105, 157
Mussolini, Benito, 89

Nagasaki, 82
Natta, Giulio, 126, 130, 143, 153
Nature, 20
Nazi Germany, 40, 48–49
Nernst, Walther, 47
New Calendar of Great Men, 135
Newton, Isaac, 58, 135–136, 144
Nielsen, Henry and Keld, 129
Nitroglycerin, 2
Nixon, Richard, 105
Nobel, Alfred, 1–4, 11–12, 35, 161
  birth and youth, 2
  death, 5
  last will, 4–7, 12, 71
  love of literature, 3
  and Mittag-Leffler, 17
Nobel, Emanuel, 6, 8–9, 12

INDEX

Nobel, Emil Oskar, 2
Nobel, Hjalmar, 6
Nobel, Ludwig, 1, 3
Nobel Brothers Petroleum Company, 8–9
Nobel committee, 62
Nobel Foundation, 6, 9
Nobel medal, 31, 33
Nobel Prizes, 9. *See also* Chemistry, Nobel Prize for; Medicine, Nobel Prize for; Physics, Nobel Prize for
  award elapsed time, 73–74
  ceremonial presentation of, 141–145
  criticism of, 14–16
  disciplinary boundaries, 51–53
  diversity of, 12
  Literature, 10, 12, 67, 141
  and nationalism, 36, 38–40, 126–128
  number of, 14, 19–20
  and Olympics, 39–41
  Peace Prize, 9, 12
  posthumous awards, 18
  prize money, 10, 31–32
  requests for, 155
  responses to, 12–14
  and science, 54, 159–162
  unsuccessful nominations, 89–92, 94–95
Nobel Prize sperm bank, 151
Nobel Prize winners
  absent, 66–71
  age, 18, 25–27, 73
  bodies of, 150–151
  characteristics of, 142–144
  citation counts, 20
  and legitimization, 119–120
  most popular, 126–132
  nationality, 19, 35–37, 129
  newspaper coverage of, 109–116
  nomination of, 27–28, 51
  peace appeals by, 105
  per year, 18
  relics of, 139–140, 152
  religion, 20
  spouses and families, 154–155
  students/collaborators of, 19–20
  third persons, 99–100
  visibility of, 130–131
  women, 18
Noble gases, 90
Noel-Baker, Philip, 40
Nordenfelt, Thorsten, 4
Nordic Assembly of Naturalists, 65
Norrby, Erling, 89, 95
Northrop, John H., 94
Norway, 6, 48
Novoselov, Konstantin, 153
Noyce, Robert, 22
Nuclear fission, 49
Nuclear research, 83
Nucleus, stability of, 25
Nurse, Paul, 32, 128, 142
Nüsslein-Volhard, Christiane, 32

Occhialini, Giuseppe, 96
Olympic Games, 39–41
Onnes, Heike Kamerlingh, 87
On the Origin of Species, 136
Osborn, Henry, 87
Oscar, king of Sweden, 9
Oseen, Carl Wilhelm, 57–59, 61–64, 66
Ossietzky, Carl von, 40, 48
Ostwald, Wilhelm, 41–43, 62

Paralytic dementia, 76
Pascal, Blaise, 104
Pasternak, Boris, 67
Pasteur, Louis, 89, 135–136, 138
Pasteur Institute, 19
The Patent Bacillus, 3
Paterniti, Michael, 140
Pauli, Wolfgang, 141
Peace appeals, 105
Peace Prize, Nobel, 9, 12
Perelli, Tommaso, 139
Periodic table of elements, 42
The Periodic Table, 90

Perutz, Max, 53–54, 95
Pettersson, Otto, 8
Peyton Rous, Francis, 26, 33, 140
Photoelectric effect, 64
Physics, Nobel Prize for
  Amano, Hiroshi, 105
  Anderson, Carl, 96
  Appleton, Edward, 55
  Ashkin, Arthur, 26
  Bardeen, John, 125
  Barkla, Charles Glover, 109
  Becquerel, Henri, 51
  Bethe, Hans, 55
  Blackett, Patrick, 96
  Bohr, Niels, 27
  Born, Max, 149
  Bragg, William and Lawrence, 26
  Brattain, Walter, 125
  Braun, Karl Ferdinand, 85
  Chandrasekhar, Subrahmanyan, 96
  Cooper, Leon, 125
  Cornell, Eric, 144
  Curie, Marie and Pierre, 27, 51
  Davis, Raymond, Jr., 25–26
  Fermi, Enrico, 27
  Feynman, Richard, 32
  Fowler, William A., 96
  Franck, James, 33
  Geim, Andre, 156
  Giacconi, Riccardo, 132
  Goeppert-Mayer, Maria, 25
  Haroche, Serge, 32
  Hess, Victor, 55
  Hewish, Antony, 96
  Jensen, Hans, 25
  Kao, Charles Kuen, 38
  Ketterle, Wolfgang, 32
  Kilby, Jack, 25
  Landau, Lev, 71
  Laue, Max von, 26
  Lawrence, Ernest Orlando, 67
  Lee, Tsung Dao, 38
  Lenard, Philipp von, 41
  Marconi, Guglielmo, 27
  Michelson, Albert Abraham, 36–37
  nominators, 27–28
  number of, 20
  Planck, Max, 46
  Powell, Cecil, 96
  Röntgen, Wilhelm Conrad, 10
  Rubbia, Carlo, 72
  Ruska, Ernst, 72
  Rutherford, Ernest, 27
  Ryle, Martin, 96
  Salam, Abdus, 128
  Schreiffer, John, 125
  Segrè, Emilio, 51
  Shockley, William, 125
  Siegbahn, Karl "Manne," 49, 80, 83, 109
  Smoot, George, 32
  Stark, Johannes, 46
  Strutt, John William (Lord Rayleigh), 104
  Ting, Samuel, 153
  Tsui, Daniel C., 38
  van der Meer, Simon, 72
  Wien, Wilhelm, 41
  Wigner, Eugene, 25
  Yang, Chen Ning, 38
Physiology, 54
Physiology or medicine, Nobel Prize for. *See* Medicine, Nobel Prize for
Pi-meson, 96
Pio XII, Pope, 89
Pizzorno, Alessandro, 116–117
Planck, Max, 13, 46–47, 63, 75, 82, 87
Platypus, 50–51
Pocketronic, 22
Poincaré, Henri, 75, 87, 91
Polanyi, John, 142
Poliomyelitis vaccines, 91
Polymerase chain reaction (PCR) technique, 20
Positron, 96
Powell, Cecil, 96
Pregl, Fritz, 32

INDEX

Prestige transfer, 125
"Prometheus Unbound," 3
Proust, Marcel, 104
Prudhomme, Sully, 10
Public lecture, Nobel Prize, 9

Rabi, Isidor, 71
Radiation therapy, 75
Radioactivity, 52, 130
Radium, 115
Ramon, Gaston, 89
Ramon y Cajal, Santiago, 140
Ramsay, William, 42
Rathenau, Walther, 63
Rayleigh, Lord, 104
Relativity, theory of, 62, 65, 117
Reppe, Walter, 90
Reputation, 116–117
Researchers, active, 73
Richards, Theodore W., 42, 47, 71
Richet, Charles, 144
Righi, Augusto, 87
Ringo Starr effect, 99
RNA Tie Club, 157
Robbins, Frederick, 91
Roberts, Richard, 32
Rock, John, 72
Rolland, Romain, 86
Röntgen, Wilhelm Conrad, 10–11, 32, 41, 106–109, 114, 116, 154
Ross, Ronald, 95–96
Rousseau, Jean-Jacques, 104
Roux, Émile, 89
Royal Society, London, 87
Royal Swedish Academy of Sciences, 8, 27–28, 42, 43, 66
Rubbia, Carlo, 72, 105, 126, 128–130, 141–142, 156
Rumford medal, 21
Ruska, Ernst, 72
Rutherford, Ernest, 20, 27, 29, 42–43, 50, 53, 56
Ruzicka, Leopold, 48

Sabin, Albert Bruce, 91
Sakharov, Andrei, 67
Salam, Abdus, 128
Salimbeni, Leonardo, 136
Salk, Jonas, 91
Sanger, Frederick, 27, 54, 125, 129
Saremo Festival, 120
Scarpa, Antonio, 139
Schreiffer, John, 125
Schrödinger, Erwin, 131
Science
 as bride, 148
 public image of, 54, 159–162
Scientific articles, 73
Scientists
 celebrity, 104–105, 114, 116, 118–119
 changing roles of, 162
 double body of, 138–139
 as experts, 118
 and humor, 123–124
 iconography of, 134–135
 media visibility, 118–119, 129
 moral exceptionality of, 146–150, 160
 narratives of, 121, 159–161
 relics of, 139
 secular worship of, 134–137
Segrè, Emilio, 51, 126, 130, 141, 154
Semenov, Nikolaj, 53
Sharpless, Barry, 125
Shelley, Percy Bysshe, 3
Shockley, William, 125, 151
Siegbahn, Manne, 49, 80, 83, 109
Silicon, 22
Sleeping sickness, 87–88
Slit lamp, 61
Smalley, Richard, 99, 103, 126
Smith, Adam, 147
Smith, Michael, 54
Smoot, George, 32
Soddy, Frederick, 56
Sohlman, Ragnar, 6–8, 10–12
Solzhenitsyn, Aleksandr, 67
Sommerfeld, Arnold, 90, 92

Sorbonne, 19
Soviet Union, 52
Speed, Odile, 92
Sperm bank, Nobel, 151
Spiro, Elfriede, 154
Sputnik effect, 148
Stadhuset, 33
Stark, Johannes, 44, 46, 62
Starr, Ringo, 99–100
Steinman, Ralph M., 29, 71
Stendhal (Marie-Henri Beyle), 104
Steptoe, Patrick, 72
Stern, Otto, 84
Stockholm engagements, 29–31
Storch, Marcus, 40
Strassmann, Fritz, 80–81
Strehlenert, R. W., 4
Strickland, Donna, 25
Strindberg, August, 11
Strutt, John William (Lord Rayleigh), 104
Superconductivity, 125
Surgery, 55
Suttner, Arthur von, 3
Suttner, Bertha von (*née* Kinsky von Wchinitz und Tettau), 3–4, 12
Svedberg, Theodor, 82–83
Sweden, 6–8, 35–36, 110
Swedish Academy, 42, 49
Synge, Richard L. M., 102
Syphilis, 76
Szilárd, Leó, 92

Teocoli, Teo, 124
Tesla, Nikola, 13, 86–87, 118
Texas Instruments, 21
Theory of Moral Sentiments, 147
Third persons, 99–100
Ting, Samuel, 153
Tiselius, Arne, 53, 94
Tissandier, Gaston, 135
Tolstoj, Lev, 11
Transistors, 21
Trapp, Yuli, 5

Tsui, Daniel C., 38
Tswett, Mikhail Semënovič, 100–102
Tuberculosis, 75
Tu Youyou, 38, 152

United Kingdom prize-winners, 19, 36
United States prize-winners, 19, 36
Urey, Harold, 105
Usmanov, Alisher, 33

Van der Meer, Simon, 72, 128
Van 't Hoff, Jacobus Henricus, 10
Veblen, Thorstein, 159
Virtanen, Artturi Ilmari, 50
Visibility, 117–118
Visible scientists, 117–118
Viviani, Vincenzo, 136
Volta, Alessandro, 136
Von Behring, Emil, 10, 89
Von Békésy, Georg, 32
Von Heidenstam, Verner, 11, 141
Von Laue, Max, 26–27, 33, 42, 50, 109
Von Lenard, Philipp, 36, 41, 44, 62, 64, 108
Von Suttner, Arthur, 3
Von Suttner, Bertha (*née* Kinsky von Wchinitz und Tettau), 3–4, 12

Wagner-Jauregg, Julius, 76
Wałęsa, Lech, 67
Walton, Ernest, 20
Wang, Jack, 33
Warburg, Otto, 94
Watson, James, 27, 33, 53–54, 92–95, 99, 105, 126, 128, 131, 146, 149–150, 157
Watts, James, 77
Wegener, Alfred, 55
Weigl, Rudolf, 90
Weller, Thomas, 91
Westgren, Arne, 82
Weyland, Paul, 62, 65
What Is Life?, 131

Widmalm, Sven, 39
Wieland, Heinrich, 47
Wien, Wilhelm, 41, 85
Wigner, Eugene, 25
Wilkins, Maurice, 53–54, 93, 95, 99, 126
Willstätter, Richard Martin, 43, 87, 101
Wireless telegraph, 111
Women, prize-winning, 18, 25
Woodward, Robert, 141
World War I, 42, 44
World War II, 49

Xenon, 90
X-rays, 10, 26, 106–109

Yang, Chen Ning, 38

Zinin, Nikolaj, 5
Zola, Émile, 104
Zuckerman, Harriet, 20
Zyklon B, 48